First International Scientific Conference on the Environment of Marshes and Water Surfaces (ISCEMWS-2021)

IOP Conference Series: Earth and Environmental Science
Volume 1002

Al-Nasiriyah, Iraq
17 – 18 November 2021

ISBN: 978-1-7138-5444-9
ISSN: 1755-1307

Printed from e-media with permission by:

Curran Associates, Inc.
57 Morehouse Lane
Red Hook, NY 12571

Some format issues inherent in the e-media version may also appear in this print version.

Institute of Physics
Dirac House, Temple Back
Bristol BS1 6BE UK

Phone: 44 1 17 929 7481
Fax: 44 1 17 920 0979

techtracking@iop.org

Additional copies of this publication are available from:

Curran Associates, Inc.
57 Morehouse Lane
Red Hook, NY 12571 USA
Phone: 845-758-0400
Fax: 845-758-2633
Email: curran@proceedings.com
Web: www.proceedings.com

TABLE OF CONTENTS

Conference Preface

Peer Review Declaration

Biodiversity of Mollusca in Bahr AL-Najaf Depression in Winter 1
A H Mizhir, N A Jasim

Study of Insect Diversity in Areas Surrounding the Central Marsh in Southern Iraq 10
Ali Kareem, Raad K. Aljaafari, Manar A. Abaas, Sienaa M. Al-Zurfi, Nadheer A. Fazaa, Omar F. Al-Sheikhly, Mukhtar K. Haba, Mark J. Whittingham

Effect of Enzyme Supplementation in Diet on Growth of Common Carp (Cyprinus Carpio L.) Juveniles in RAS 19
Noori Abdul-Nabi Nasir, Abdulkareem Taher Yesser, Qusay Hamid Al-Hamadany

Desalination Methods and Their Role in Solving and Managing the Problem of Salinization in Iraqi Water Resources 30
Havan H. Salman, Zahraa S. Aswad

Study Kinetic Reaction and Removal of Indigo Carmine Dye in Aqueous Solutions by Direct Electrochemical Oxidation 51
Karrar H. Kazm, Sarmad T. Najim

Cadmium Resistant Bacteria Mediated Cadmium Removal: A Systematic Review on Resistance, Mechanism and Bioremediation Approaches 65
Nadia Jebril, Rich Boden, Charlotte Braungardt

Solid Waste: A Study of Its Concept, Management Methods, and Environmental Impacts 89
Suaad Hadi Hassan Al-Taai

Air Pollution Resulted from Coal, Oil and Gas Firing in Thermal Power Plants and Treatment: A Review 104
Rafid M. Hannun, Ali H. Abdul Razzaq

Water Management Scheme to Restore and Sustain the Marshes and Shatt al-Arab, Southern Iraq........113
Khayyun Rahi, Ali Abdulkhabeer Ali

Analysis of the Environmental Reality of the Marshes and Its Sustainable Development........ 121
Israa Mahmooed Mohammed Irzoqy, Lamia Flaieh Ibrahim, Hodoud Mohamed Abboud Al-Tufaily

Effluent Quality Assessment Monitoring of Al-Rustamiya Sewage Treatment Plant Using Geographical Information System (GIS) 139
Rusol Mohammed, Basim Huseen, Athraa Hashim

Use Remote Sensing Techniques to Study Epiphytic Algae on Phragmites Australis in Um El-Naaj Lake, Mysan Province, Southern Iraq 154
Al Naqeeb Neran A., Al Hassany Jinan S., Mashee Fouad K.

The Effect of Climate Change on Energy Consumption and Environmental Pollution According to Degree Days Indicator "Baghdad City as an Example" 164
Susan Abed Hassan, Saad Ismail Husain

Author Index

IOP Conf. Series: Earth and Environmental Science 1002 (2022) 011001 doi:10.1088/1755-1315/1002/1/011001

Conference preface

First international conference of marshes and water surfaces

The conference execution: Actually (Opening session in the first day) & virtually

Reason: COVID-19 and travel constriction.

Technology used for conference delivery: On line via "ZOOM".

Date of conference: Wednesday and Thursday Nov. 17th 18th - 2021

Location of organization: Marshes Research Center /University of Thi-Qar, Nasryiah, Thi-Qar, Iraq

Presentation type: invited, contributed.

Conference session: two

Topic or subject area:

1- Biodiversity, Conservation in Marshes Environment.

2- Water and Wastewater Treatment.

3- Environmental Effect on Health.

4- Environmental Sustainability.

5- Water Resources Management and Remote Sensing.

6- Earth and Sediment Sciences

Sponsor: Participates (Author(s) of paper)

Funding: non (conference participation fees)

Acknowledgements: Many thanks Dr. Mohammed Muayad TA /Environment Research Center, University of Technology-Iraq and University of Thi-Qar for regarding protocols of conference holding.

IOP Conf. Series: Earth and Environmental Science 1002 (2022) 011001 doi:10.1088/1755-1315/1002/1/011001

Lists of committees:

1- Supreme Supervising Committee

- Prof.Dr. Yahya Abdul-Redha Abbas Alkhafaji
- Prof Dr.Kamal H.Yasir (Assistant of University President for Scientific Affairs)
- Assist. Prof. Dr. Ali Abdulkhabeer Ali (Director of Marshes Research Center)

2- Organizing Committee

- Assist Prof Dr. Ali Abdulkhabeer Ali
- Assist Prof. Dr. Amjad Hameed Al-mudhafer, Iran
- Dr. Hussein A. H. Al-Sa'idy
- Dr. Khudhair Joudah Yasir
- Dr. Omar Fadhil Al sheikhly
- Dr. Azhar Abdulabass Jaffar
- Dr.Sadiq K. Odeh
- Dr. Raed Kadhum Fakher Alfahed
- Falah Kareem Hadi Al-Kaabi
- Dr. Zaman kareem hanan
- Basim Turki Alyousif
- Ali Hussain Lazem
- Sadiq Hassan Jassim
- Irfan Jasour Musa
- Ahmed Hussam Abdel Hamid
- Hanan Mohammed Saleh Salman
- Ban Naji Hamdan

3- Scientific Committee

- Prof. Dr. Abdul Hameed M. Jawad Al Obaidy, Iraq
- Prof. Dr. Basim Yousif Dhiab Alkhafaji, Iraq
- Prof. Dr. Ahmed Shaker alashoor, Iraq
- Prof. Dr. Rafid M. Hannun, Iraq
- Prof. Dr. Razzaq Ghazi Neghamish, Iraq
- Prof. Dr. Raheem Hameed Alabdan, Iraq
- Prof. Dr. Ali Abd Saadoon Al.Ghuzi, Iraq
- Prof. Dr. Noori Abdul- nabi Nasir, Iraq
- Prof. Dr. Amjed Kadhum Resen, Iraq
- Prof. Dr. Khalid Mohammed Hussein Al-Qaisi, Iraq
- Prof. Dr. Nihad Abdul-Lateef Ali, Iraq
- Prof. Sadek Jafar Tuama, Iraq
- Prof. Dr. Mehdi Ghomeshi, Iran
- Prof. Dr. Seyed Mahmood Kashefipour, Iran
- Prof. Dr. Mark Van Loosd Vecht, Netherlands
- Prof. Dr. Elena Romenovna Magaril, Russia
- Prof .Dr. Nadhir Al-Ansari, Sweeden
- Prof. Dr. Gamal Gomaa Medani, Egypt
- Prof. Dr. Adnan Ali Nizam, Syria
- Assist. Prof. Dr. Fadhil Sarhan Kadhim , Iraq
- Assist. Prof. Dr. Najim A. Al-Rubaye, Iraq
- Assist. Prof. Dr. Fathi Almandeel, Iraq
- Assist. Prof. Dr. Salam Hussein Ewaid, Iraq
- Assist. Prof. Dr. Saleem Ethaib Mohammad, Iraq
- Assist. Prof. Dr. Mohammed A. Alyasiri, Iraq
- Assist. Prof. Dr. Mohammad Mahmoud Farhan Al-Halbosiy,Iraq
- Assist. Prof. Dr. Mukhtar K. Haba, Iraq
- Assist. Prof. Dr. Kazem Rangzan, Iran
- Assist. Prof. Dr. Khayyun A. Rahi, USA
- Assist. Prof. Dr. Wasseem Morshad, Syria

Participants:

Topic (1) Biodiversity, Conservation in Marshes Environment				
Authors	**ID**	**Title**	**Email**	**Program**
- Aliaa Hussein Mizhir - Nahdema Abdabbas Jasim	001_ISCEMWS	Biodiversity of Mollusca in Bahr Al-Najaf Depression in Winter	aliah.alamary@uokufa.edu.iq	Zoom
- Ali Kareem - Manar A. Abaas - Raad K. Aljaafari - Nadheer A. Fazaa	002_ISCEMWS	Study of insect diversity in The Central Marsh in southern Iraq	ali.kareem@uokerbala.edu.iq	Zoom
- Noori Abdul-nabi Nasir - AbdulKareem Taher Yesser - Qusay Hamid Al-Hamadany	003_ISCEMWS	Effect of Enzyme Supplementation in Diet on the Growth of Common Carp (Cyprinus carpio L.) juveniles in close system	noori.nasir@uobasrah.edu.iq	Zoom
Topic (2) Water and Wastewater Treatment				
Authors	**ID**	**Tittle**	**Email**	**Program**
- Dr. Havan H. Salmant - Zahraa S. Aswad	004_ISCEMWS	Desalination methods and their role in solving and managing the problem of salinization in Iraqi water resources	havan.hassan68@gmail.com	Zoom
- Karrar H. Kazm - Sarmad T. Najim	005_ISCEMWS	Study Kinetic Reaction and Removal of Indigo Carmine Dye in Aqueous Solutions by Direct Electrochemical Oxidation	karrarhaider177@yahoo.com	Zoom
- Nadia Jebril - Rich Boden - Charlotte Braungardt	006_ISCEMWS	Cadmium resistant bacteria mediated cadmium removal: a systematic review on resistance, mechanism and bioremediation approaches	nadia.tawfiq@uobabylon.edu.iq	Zoom
Topic (3) Environmental Effect on Health				
Authors	**ID**	**Tittle**	**Email**	**program**
- Suaad Hadi Hassan Al-Taai	007_ISCEMWS	Solid waste: A study of its concept, management methods, and environmental impacts	suaad.hadi@ircoedu.uobaghdad.edu.iq	Zoom
- Rafid M. Hannun - Ali H. Abdul Razzaq	008_ISCEMWS	Air Pollution Resulted from Coal, Oil and Gas Firing in Thermal Power Plants and Treatment: A Review	Rafid-m@utq.edu.iq	Zoom
Topic (4) Environmental Sustainability				
Authors	**ID**	**Tittle**	**Email**	**Program**
Khayyun Rahi Ali Abdulkhabeer Ali	009_ISCEMWS	Water management scheme to restore and sustain the Marshes and Shatt al-Arab, Southern Iraq	khayyun.rahi@okstate.edu Aranru79@utq.edu.iq	Zoom

IOP Conf. Series: Earth and Environmental Science 1002 (2022) 011001 doi:10.1088/1755-1315/1002/1/011001

- Susan Abed Hassan -Saad Ismail Husain	010_ISCEMWS	The effect of climate change on energy consumption and environmental pollution according to degree days indicator "Baghdad city as an example"	dr.susanabedhassan@eng.nahrainuniv.edu.iq	Zoom
- Israa Mahmooed mohammed Irzoqy - Hodoud Mohamed Abboud Al-Tufaily	011_ISCEMWS	Analysis of the environmental reality of the marshes and its sustainable development	Znbg50544@yahoo.com	Zoom
Topic (5) Water Resources Management and Remote Sensing.				
Authors	**ID**	**Tittle**	**Email**	**Program**
- Rusol Mohammed Mohsen - Basim Huseen - Athraa Hashim	012_ISCEMWS	Effluent Quality Assessment Monitoring of Al-Rustamiya Sewage Treatment Plant Using Geographical Information System (GIS)	r.mohsin1901m@coeng.uobaghdad.edu.iq	Zoom
- AL Naqeeb, Neran A. - AL Hassany, Jinan S. - Mashee, Fouad K.	013_ISCEMWS	Use Remote Sensing Techniques to Study Epiphytic Algae on Phragmites australis in Um El-Naaj Lake, Mysan Province, Southern Iraq	neranecology@uomisan.edu.iq	Zoom

Time of presentation for each participant: 15 min. (presentation & discussion).

Questions and discussions:

One of the remarkable discussions issue was the influences of climate changes and foreign political affairs regarding water rations among neighboring countries on submerged spaces of Iraqi Marshes.

IOP Conf. Series: Earth and Environmental Science 1002 (2022) 011001 doi:10.1088/1755-1315/1002/1/011001

Conference photographs:

IOP Conf. Series: Earth and Environmental Science 1002 (2022) 011001 doi:10.1088/1755-1315/1002/1/011001

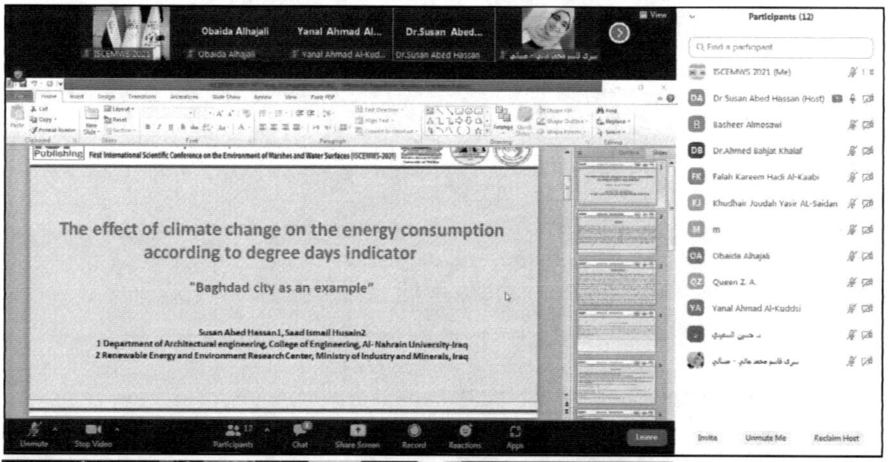
Participants (12)
Obaida Alhajali
Yanal Ahmad Al...
Dr.Susan Abed...
First International Scientific Conference on the Environment of Marshes and Water Surfaces (ISCEMWS-2021)
The effect of climate change on the energy consumption according to degree days indicator
"Baghdad city as an example"
Susan Abed Hassan1, Saad Ismail Husain2
1 Department of Architectural engineering, College of Engineering, Al- Nahrain University-Iraq
2 Renewable Energy and Environment Research Center, Ministry of Industry and Minerals, Iraq
ISCEMWS 2021 (Me)
Dr Susan Abed Hassan (Host)
Basheer Almosawi
Dr.Ahmed Bahjat Khalaf
Falah Kareem Hadi Al-Kaabi
Khudhair Joudah Yasir AL-Saidan
Obaida Alhajali
Queen Z. A.
Yanal Ahmad Al-Kuddsi
Unmute
Stop Video
Participants
Chat
Share Screen
Record
Reactions
Apps
Leave
Invite
Unmute Me
Reclaim Host

Participants (18)
Dr. Ali Abdulkh...
Yanal Ahmad Al...
ISCEMWS 2021 (Me)
Obaida Alhajali (Host)
Dr. Ali Abdulkhabeer Ali
Ghada Bilal
Yanal Ahmad Al-Kuddsi
dr. khalid AL-qaysi
Dr.Ahmed Bahjat Khalaf
Dr.Susan Abed Hassan
Falah Kareem Hadi Al-Kaabi
Khudhair Joudah Yasir AL-Saidan
Queen Z. A.
Sabah Abbas Malik
Unmute
Start Video
Participants
Chat
Share Screen
Record
Reactions
Apps
Leave
Invite
Unmute Me
Reclaim Host

IOP Conf. Series: Earth and Environmental Science 1002 (2022) 011001 doi:10.1088/1755-1315/1002/1/011001

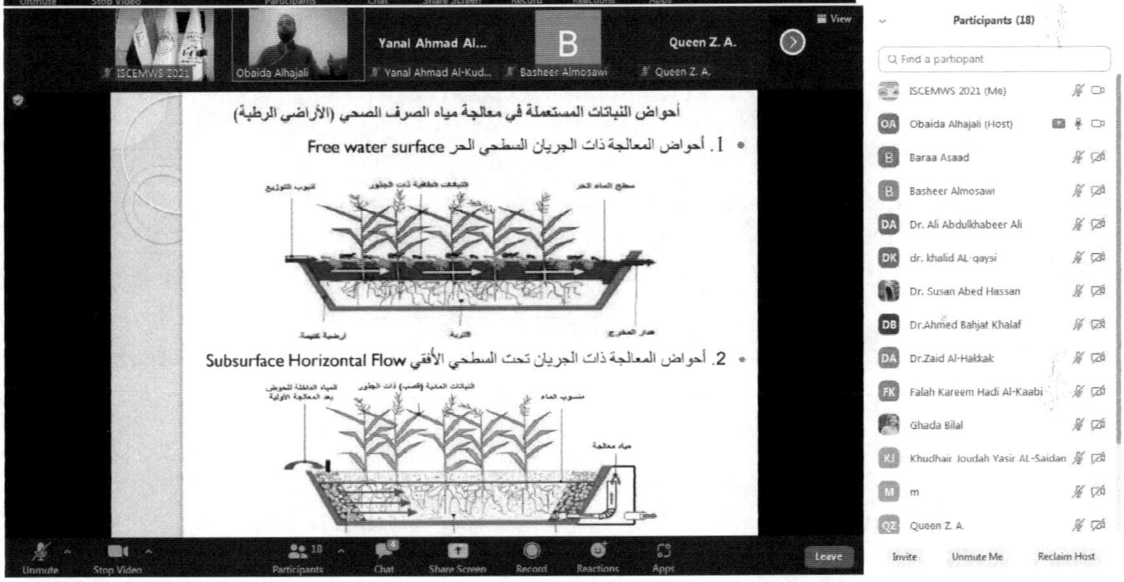

IOP Conf. Series: Earth and Environmental Science 1002 (2022) 011001 doi:10.1088/1755-1315/1002/1/011001

001_ConfName.pdf
002_ConfName.pdf
003_ConfName.pd

Peer Review Declaration

Type of peer review: double blind

Conference submission management: Via Google form and e-mails

con.utq.edu.iq

https://forms.gle/mvgcXWUKRkdgZjMw7

isce2021.mrc@utq.edu.iq

ISCEMWS1@utq.edu.iq

ISCEMWS2@utq.edu.iq

ISCEMWS3@utq.edu.iq

ISCEMWS4@utq.edu.iq

ISCEMWS5@utq.edu.iq

IOP Conf. Series: Earth and Environmental Science 1002 (2022) 011001 doi:10.1088/1755-1315/1002/1/011001

Overall participant number: (13 papers presenters + 85 attendance without fees)

Number of submission received: 42

Number of submission sent for review: 35

Number of submissions accepted: 13

Total number of reviewers involved: 82

Acceptance rate:100 * (13/35) = 37%

- The low acceptance rate is because of the following reasons
  A- This conference is actually our first international we were expecting low participation profile because its (the conference) is not glow body famous yet.
  B- Some of the low scientific quality papers have been rejected by the reviewers.
  C- Some of the authors have withdrawn their papers from participation after peer reviewing because of unknown reasons
  D- Some other authors have failed to deliver their papers after correction in the declared dead time

Average number of reviewer per paper: Two - Three

Any additional information about review process: check of grammar and plagiarism

Important note: Plagiarism of paper No. 006_ISCEMWS is 90% because it's part of them thesis.

Contact person for queries:

Dr. Ali Abdulkhabeer Ali

Director of Marshes Research Center/ University of Thi-Qar/Iraq

Aranru79@utq.edu.iq

IOP Conf. Series: Earth and Environmental Science 1002 (2022) 011002 doi:10.1088/1755-1315/1002/1/011002

Peer Review Declaration

All papers published in this volume of IOP Conference Series: Earth and Environmental Science have been peer reviewed through processes administered by the Editors. Reviews were conducted by expert referees to the professional and scientific standards expected of a proceedings journal published by IOP Publishing.

Type of peer review: Double blind
Conference submission management: Via Google form and e-mails
con.utq.edu.iq
https://forms.gle/mvgcXWUKRkdgZjMw7
isce2021.mrc@utq.edu.iq
ISCEMWS1@utq.edu.iq
ISCEMWS2@utq.edu.iq
ISCEMWS3@utq.edu.iq
ISCEMWS4@utq.edu.iq
ISCEMWS5@utq.edu.iq

Overall participant number: (13 papers presenters + 85 attendance without fees)

Number of submissions received: 42

Number of submissions sent for review: 35

Number of submissions accepted: 13

Total number of reviewers involved :82

Acceptance rate:100 *(13/35) = 37%

- The low acceptance rate is because of the following reasons
  - A- This conference is actually our first international we were expecting low participation profile because its (the conference) is not glow body famous yet.
  - B- Some of the low scientific quality papers have been rejected by the reviewers.
  - C- Some of the authors have withdrawn their papers from participation after peer reviewing because of unknown reasons
  - D- Some other authors have failed to deliver their papers after correction in the declared dead time

Average number of reviewers per paper: Two- Three

Any additional information about review process: check of grammar and plagiarism
Important note: Plagiarism of paper No. 006_ISCEMWS is 90% because it's part of them thesis.

Contact person for queries:
Dr. Ali Abdulkhabeer Ali
Director of Marshes Research Center/ University of Thi-Qar/Iraq
Aranru79@utq.edu.iq

IOP Conf. Series: Earth and Environmental Science 1002 (2022) 012001 doi:10.1088/1755-1315/1002/1/012001

Biodiversity of Mollusca in Bahr AL-Najaf Depression in winter

A H Mizhir and N A Jasim

Ecology and Pollution Department / Faculty of Science /University of Kufa . Najaf \ Iraq

aliah.alamary@uokufa.edu.iq

Abstract. To study the biological diversity of the Mollusca, six sites were selected along the perimeter of the Bahr AL-Najaf Depression, and samples were collected at a rate of three replicates monthly for the interval from November 2019 till January 2020. The results of this study showed that highest density of 224 individuals / 50 cm^2 was recorded in January 2020. Eight species have been identified, 4 of which belong to Bivalvia: *Corbicula fluminea*, *Pseudodontopsis euphraticus*, *Unio tigridis* and *Trachycardium lacunosum*, 4 species belong to the gastropod: *Melanopsis nodosa*, *Melanopsis costata*, *Melanoides tuberculata* and *Theodoxus jordani*. *Corbicula fluminea* and *Melanoides tuberculata* were more abundant in study site compared with else species. The results of the constancy index showed that five species, one of theme belongs to Bivalvia and the rest belongs to Gastropod, where the other species were accessories to Bahr AL-Najaf Depression environment. The index values of Species richness varied between 2.5-6.5, while the Shanon-Weiner index ranged between 0.024-0.29 bit/ind., but the uniformity index ranged between 0.024- 0.036.

1. Introduction:

Mollusca is second only to Arthropoda in numerical abundance also is the second most diverse of all animal phyla, occurring in numerous territory types [1,2,3,4,5]. Mollusca are normally found in Aplacopora, Caudofoveata, Monoplacopora, Polyplacophora, Cephalopoda, Scaphoda, Gastropoda and Bivalve. The two biggest classes of Mollusca phylum are Bivalve and Gastropoda [6]. Gastropods and Bivalves are a main ingredient of marine, coastal and fresh water, have a significant ecological role to play in their ecosystem by participating in filtration and bioturbation in lotic and lentic climates [7–9]. Living and their dead cortexes of Mollusca can trigger food accessibility and supply physical build for other organisms [10]. Ecologically, Mollusca with participation of other organisms play an important labor in stabilizing food web [11], and economically, It is an additional source of protein and increases the financial imports [12,13].

The dispersion and bounty of Mollusca are significant variables in biological administration [14]. As Molluscas biodiversity and underlying modality are perceptibly affected by: Physical and Chemical circumstances, Supplement supply and Food Web dynamics of ecological systems [9]. Mollusca body growth affecting by changing in climatic conditions [15].Hibernation and aestivation of many Mollusca species are likely occur in extreme temperature and humidity. In such manner, an inclusive comprehension of the connections between Molluscas communities and their ecological systems conditions is a significant element to support biodiversity, protection and ecological systems administration [16].

Assessment of biodiversity of particular groups and territories is the high priority task these days, and as Bahr AL-Najaf depression creates unified ecosystem, attention must be taken to a diversity of Mollusca on it. So that, current study was conducted to diagnosing Mollusca species , and applying biodiversity laws, incorporate Relative abundance index (Ra), Constancy index (S), species richness index (D), Shannon-Weiner diversity index (H), and uniformity index (E) to envisage wellbeing of waters in Bahr AL- Najaf depression

2. Materials and Methods:

The current study involves identification and quantitative analysis of Mollusca in Bahr AL-Najaf depression, southwest of AL-Najaf city/Iraq. Samples of Mollusca were collected from six sites were selected along the perimeter of Bahr AL-Najaf Depression during interval between November 2019 - January 2020. Bahr AL-Najaf depression is situated between latitudes 44.15°-44.21°N and longitudes 31.50°-32.50°E. Sampling sites are illustrated in Figure 1 and the study area include Location I- locally known as slot and situated in the end of orchard in eastern of depression(32°00'01.1"N, 44°16'49.3"E), Location II- locally known as tongue (32°01'24.6"N, 44°14'53.2"E), Location III- situate obverse Al-Hawli street (32°02'40.4"N, 44°11'49.7"E), Location IV- situate versus crematorium western Bahr AL-Najaf (32°00'39.9"N, 44°12'28.6"E), Location V- locally known as Al-Tar and situated in adjacent to main street (31°58'46.4"N, 44°13'22.4"E), Location VI- After fifth site (31°57'57.9"N, 44°15'25.4"E).

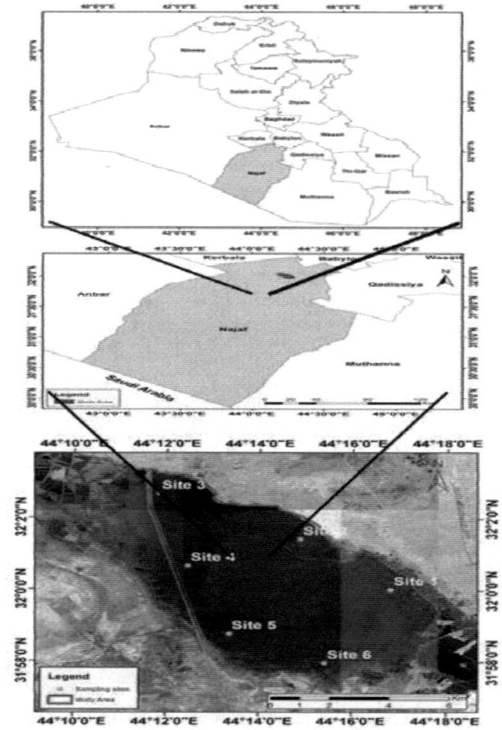

Figure (1) topographical locations from which samples were gathered from Bahr al-Najaf depression

Samples of Mollusca were collected once monthly from six sites. Quantitative analyses of the samples were carried out by collecting the specimens by the method of handpicking using a quadrate of known size. Further, Bivalves and Gastropod were collected by hand digging the substratum or by handpicking in different places of each sampling region [17]. Also individual of collected samples were sent to Marine Sciences Center / University of Basrah to diagnose.

Five biological indicators were applied on studied samples as following:

2.1. Relative abundance index (Ra): Was calculated through the subaltern equation putted by [18].

2.2. Constancy Index (S): This index enables determining the frequency occurrence of a given species, [19].

2.3. Shannon-Weiner Diversity Index (H): Was monthly calculated by the equation of Shannon-Weiner as demonstrate in [20].

< Low Diversity, (–) Moderate Diversity, > High Diversity [21].

2.4. Species Richness Index (D): was calculated from [22].

2.5. Species uniformity index(E): Was measured according to [23].

3. Results:

3.1 Physical and chemical features of water in study areas:

Water temperature values of Bahr AL-Najaf depression in the study sites ranged between (12.5–16.7)°C Figure 2. While the value of the pH ranged between (7.3–9.28) Figure 3, the value of the electrical conductivity of water ranged between (15.6–9.7) ms/cm Figure 4. Likewise, salinity values of the water ranged between (10.6–38.2) ppt Figure 5, while value of dissolved oxygen in the water was between (*2 – 3.1*) mg/L Figure 6. The values of BOD in the water varied between (0.5 –2.3) mg/L during the study months Figure 7, and nitrate values were recorded and were of varying proportions ranging from (0.007 –1.2)μg /L Figure 8.

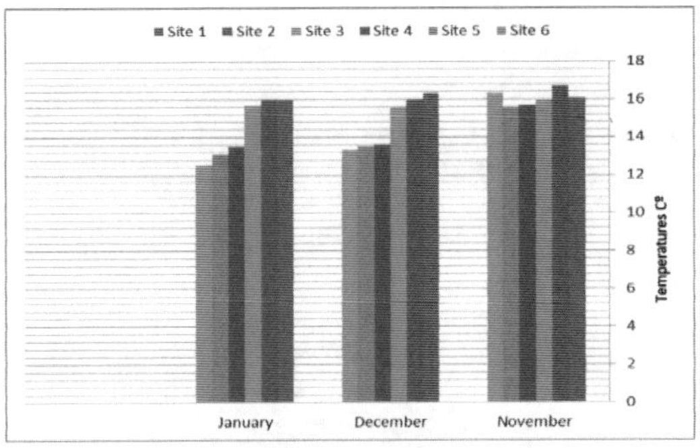

Figure (2)Water temperatures of the studied sites in Bahr AL-Najaf depression from November 2019 to January 2020.

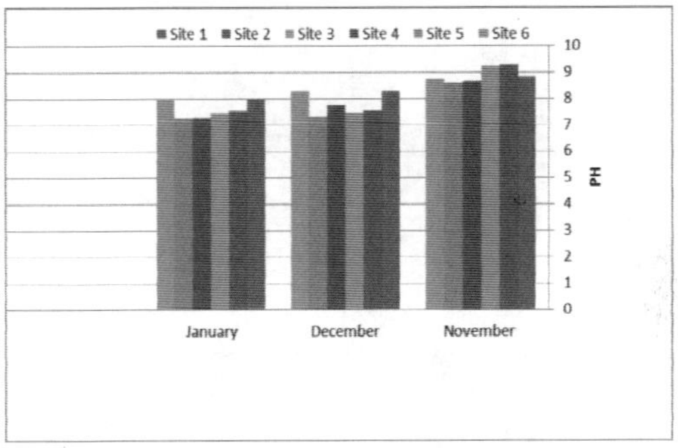

Figure (3) pH values for the studied sites in Bahr AL-Najaf depression from November 2019 to January 2020.

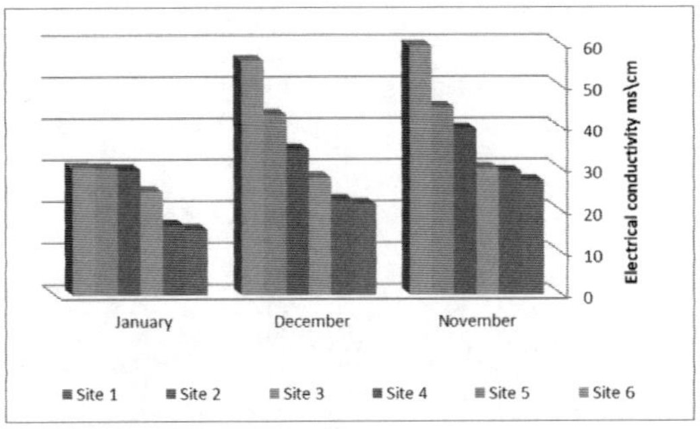

Figure (4) electrical conductivity of study sites in Bahr AL-Najaf depression from November 2019 to January 2020.

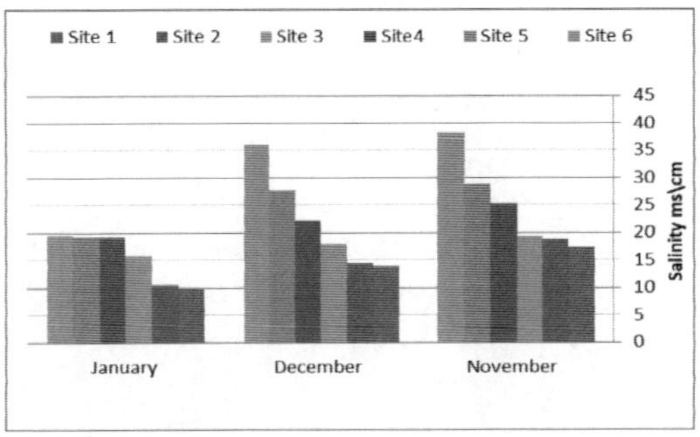

Figure (5) Salinity in study sites of Bahr AL-Najaf depression from November 2019 to January 2020.

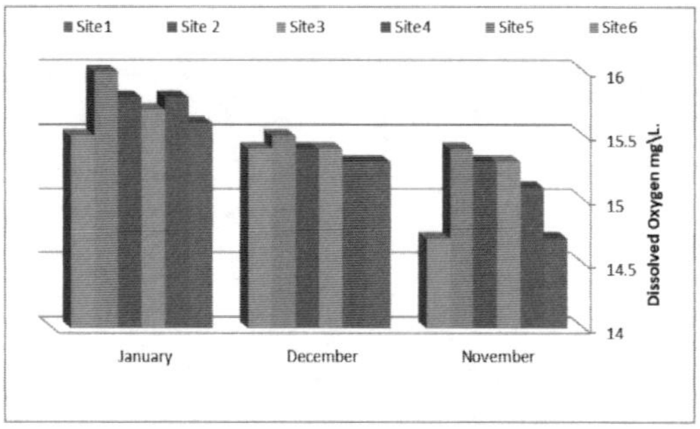

Figure (6) dissolved oxygen of water samples for study sites in Bahr AL-Najaf depression from November 2019 to January 2020.

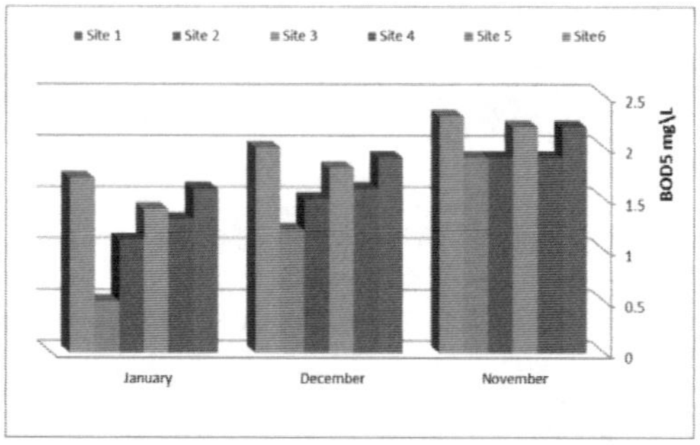

Figure (7) BOD of water samples for study sites in Bahr AL-Najaf depression from November 2019 to January 2020.

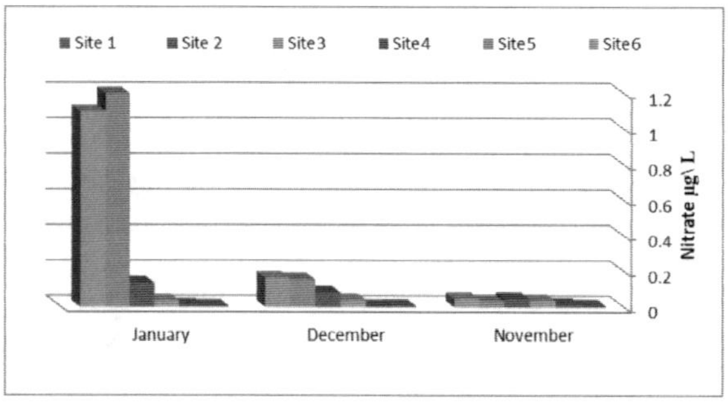

Figure (8) nitrate concentrations of water samples for the study sites in Bahr AL-Najaf depression from November 2019 to January 2020.

3.2 Density and biomarkers:

Eight species of Mollusca have been identified, 4 of which belonging to Bivalvia: *Corbicula fluminea*, *Pseudodontopsis euphraticus*, *Unio tigridis* and *Trachycardium lacunosum* Figure 9, Four species belong to the gastropod: *Melanopsis nodosa*, *Melanopsis costata*, *Melanoides tuberculata* and *Theodoxus jordani* Figure 10. Its total density ranged between 7-224 ind./50 cm^2 and the highest density was for *Corbicula fluminea*, where the highest density was in December 2019 as it reached 99 ind./50 cm^2 in the first site (Figure 11 and 12).

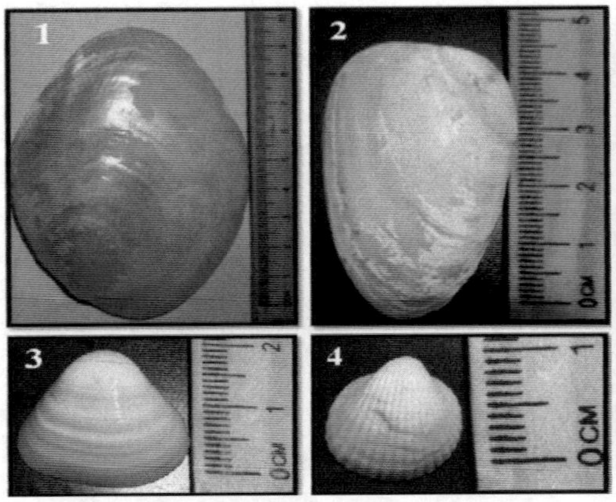

Figure (9) Bivalvia isolated from the study sites in Bahr AL-Najaf depression from November 2019 to January 2020.
1- *Pseudodontopsis euphraticus* 2-*Unio tigridis*
3- *Corbicula fluminea* 4- *Trachycardium lacunosum*

Figure (10) Gastropod isolated from the study sites in Bahr AL-Najaf depression from November 2019 to January 2020.
1- *Melanoides tuberculata* 2- *Melanopsis costata*
3- *Melanopsis nodosa* 4- *Theodoxus jordani*

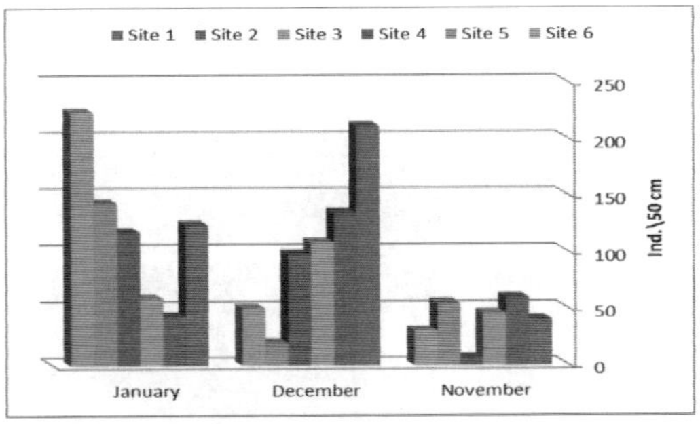

Figure (11) Total density of Mollusca in study sites of Bahr A Najaf depression from November 2019 to January 2020.

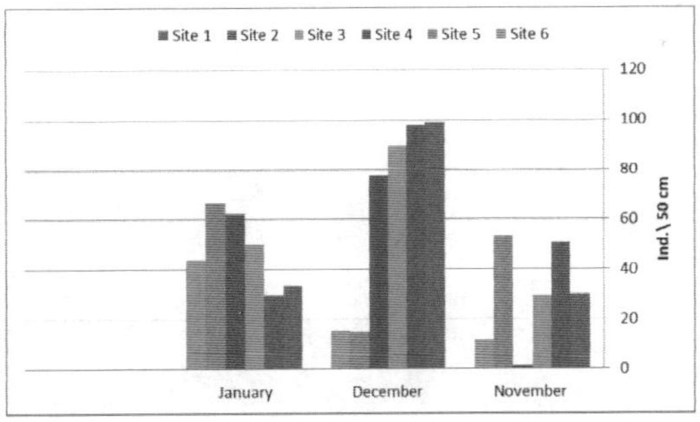

Figure (12) Total density of *Corbicula fluminea* in study sites of Bahr AL-Najaf depression from November 2019 to January 2020.

The results revealed that *Corbicula fluminea* was the most abundant species in the first site, 79%, and then *Melnnoipis nodosa* which recorded 13%. In the second site, *Corbicula fluminea* was the most abundant, with a percentage of 82%, and then *Melnnoipis nodosa*, which recorded a 7%. In the third site, *Corbicula fluminea* was the most abundant species 85%, and then *Melnnoipis costata* 6%, and in the fourth site, *Corbicula fluminea* was the most abundant 71%, and then *Melnnoipis costata* 14%, and in the fifth site, *Corbicula fluminea* recorded 64%, followed by *Melanoides tuberculate* which recorded 16%. In the sixth position, *Melanoides tuberculate* was the most abundant 39%, and then *Corbicula fluminea* 25%, followed by *Melnnoipis costata* 14% (Figure 13).

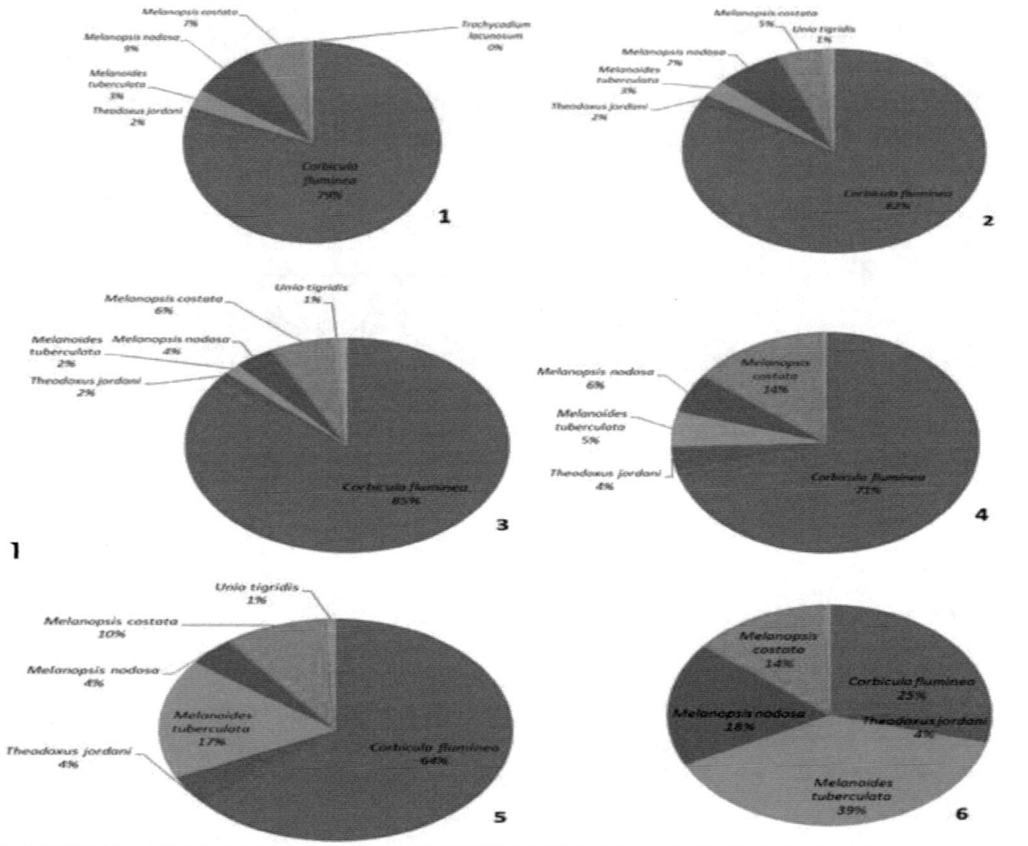

Figure (13) Indicator of relative abundance of Mollusca in study sites of Bahr AL-Najaf depression for the period from November 2019 to January 2020.

Table 1 shows constancy index which indicated that each of *Carbicula flumineu*, *Theodoxus jordani*, *Melanoides tuberculate*, *Melnnoipis nodos* and *Melnnoipis costata* were the most stable in six locations of Bahr AL-Najaf depression during three months. Furthermore, species richness index values of Mollusca ranged between lowest value (2.5) in January in sixth location and the highest value (6.5) in November in the fourth location Figure (14).

Table (1) Constancy Index of Mollusca species and their recurrence in Bahr AL- Najaf depression in winter

Study Sites						Species
Site 6	Site 5	Site 4	Site 3	Site 2	Site 1	Mollusca
						Corbicula fluminea
						Theodoxus jordani
						Melanoides tuberculata
						Melanopsis nodosa
						Melanopsis costata
						Unio tigridis
						Pseudodantopsis euphraticus
						Trachycadium lacunosum
not recorded			Constant species			additive species

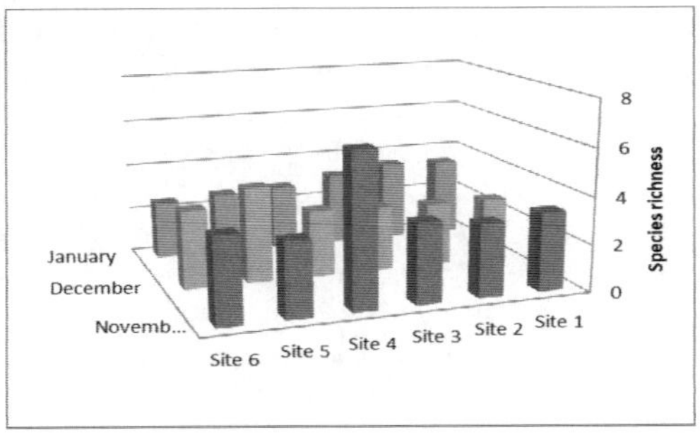

Figure (14): Species Richness index of Mollusca in study sites in Bahr AL-Najaf depression in winter

Shannon -Wiener Index detect variation from lowest value 0.024 bit/ind in January for the sixth site and the highest value 0.29 bit/ind for the fourth site in December (Figure 15). Finally, species uniformity index of Mollusca ranged between highest valued (0.036) in fourth site of December and lowest valued (0.024) in sixth site of January the as in Figure (16).

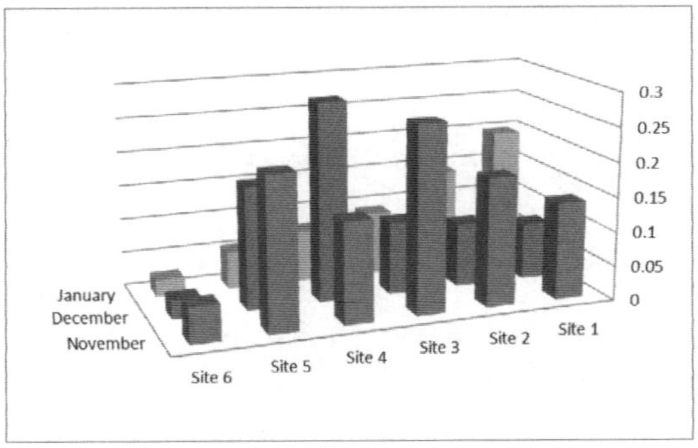

Figure (15): Shannon-Weiner index of Mollusca in different study sites in Bahr AL-Najaf depression in winter

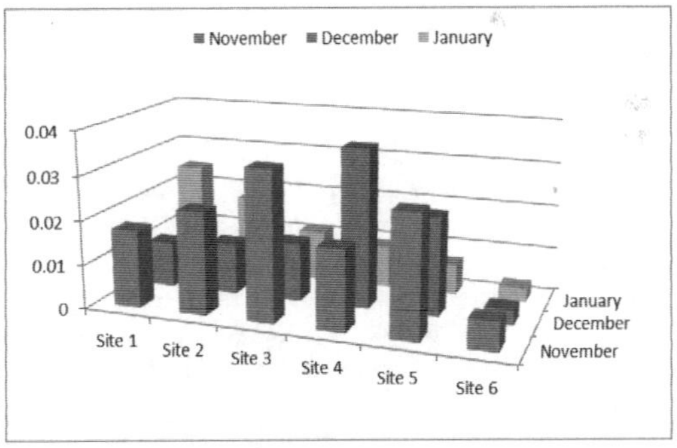

Figure (16): Species uniformity index of Mollusca in study sites in Bahr AL-Najaf depression in winter

4. Discussion:

Existing study revealed decrease in total density of Mollusca (7-224 ind./50 cm^2) in the Bahr AL-Najaf depression compared to previous Iraqi studies in different water bodies. [24] recorded the highest density of Mollusca 23,177.64 ind/m^2 and 18046.71 ind/m^2 in Tigris and Diyala rivers, respectively, and this decrease may be due to the influence of environmental factors on the density of Mollusca, as [24] indicated that there is a positive correlation between the two streams and each of the water temperature and the concentration of nitrates from nutrients. The reason for this may be due to

the Mollusca having large ranges of tolerance to environmental conditions [25, 26]. This study agrees with the study of [27], which recorded the predominance of the gastropod cultivar over other invertebrates in intertidal zone of Shatt Al-Arab.
Inability to reach high values of abundance index which shown in existing study explain inability of the dominant species to attain complete dominance over else species which gives evidence that there are no environmental stresses in Bahr AL-Najaf depression that allow of certain species to bear these stresses and achieve complete dominance on else types as indicated by [28]. Recording of low values for species richness index of Mollusca in most studied sites of Bahr AL-Najaf depression during study time was due to same reasons that led to the decrease in the density.
Furthermore, enrolment the low values of Shannon Wiener index less than 1 bit/ind. in this study was counted as less diversity. The present study also listed 6 species of Mollusca, this is different from [29] whose recorded 8 species belonging to the Molluscas class, while [30] recorded approximately 7.819 ind./bits, evidence from presence of ecological stress and high contamination, causing demise or decampment of susceptible species. So [31] refer that waters of Bahr AL- Najaf depression were polluted with some heavy metals, which may be demonstrate the reason of Shannon Wiener index reduce in the present study. Concerning species uniformity index, the results demonstrate that Mollusca were enumerated as varied in their occurrence, as uniformity values were less than 0.5, also this is additional proof of ecological stress in Bahr AL- Najaf depression. The cause of ecological pressures and the deficiency of Mollusca might be because of the presence of water currents in Bahr AL- Najaf depression and this is agree with what [23] refer to constant water leads to set of factors that help enrichment of nutrient and thus the boom of Mollusca. [25] Also refer to reduction in ecological conditions fundamental for natural variety of the organisms on which macro-invertebrates feed that may participate in reducing biodiversity index values in those districts. Some Iraqi researches also registered high abundance of macro-invertebrates during spring and autumn, and decrease their abundance during winter, as a result of combination effect of ecological conditions with the suitable time for mating and laying eggs [32, 33].

5. Conclusions

From the results obtained we conclude that the Mollusca and consequently the rest of the aquatic organisms live under environmental pressures resulting from pollution and various human activities in addition to climatic changes and the low water level in the region which requires more attention from the local government to preserve the environment of the depression and the organisms living in it.

References

[1] Morton J E, Molluscs. 4th Edition. (Hutchinson University Library. London), 1967: pp. 244.
[2] Russell-Hunter W D 1983 Overview: Planetary Distribution of and Ecological Constraints upon the Mollusca, *The Mollusca*. Wilbur K M (ed.) *Academic Press. New York* **6** 1-27.
[3] Cain A J 1983 Ecology and Ecogenetics of Terrestrial Molluscan Populations. In: *The Mollusca*. K. M. Wilbur (ed.) *Academic Press. New York* **6** 597-647.
[4] Solem A A 1984 World model of land snail diversity and abundance. In: *World-Wide Snails. Biogeographical studies on non-marine Mollusca* Solem A and van Bruggen A C (eds.) Brill/Backhuys Leiden 6-22.
[5] Shanmugam A and Vairamani S 2005 Molluscs in Mangroves *A Case Study Centre of Advanced Study in Marine Biology Annamalai University* 371 p
[6] Dharma B 2005 Recent and fossil Indonesian shells *Conch Books Hackenheim Germany* 424 pp.
[7] Bogan A E 2008 Global diversity of freshwater mussels (Mollusca, Bivalvia) in freshwater *Hydrobiologia* **595** 139–147 [CrossRef]
[8] Palmer M Covich A Finlay B Gibert J Hyde K Johnson R Kairesalo T Lake S Lovell C Naiman R *et al.* 1997 Biodiversity and ecosystem processes in freshwater sediments *Ambio* **26** 571–577.
[9] Covich A P Palmer M A Crowl T A 1999 The role in of species invertebrate freshwater ecosystems—Zoobenthic species influence energy flows and nutrient cycling *Bioscience* **49** 119–127. [CrossRef]
[10] Vaughn C C Nichols S J Spooner D E 2008 Community and foodweb ecology of freshwater mussels *J N Am Benthol Soc* **27** 409–423. [CrossRef]

[11] Green R H 1971 A multivariate statistical approach to hutchinsonian niche: Bivalve molluscs of Central Canada *Ecol* **52** 544–556. [CrossRef] [PubMed]

[12] Köhler F Seddon M Bogan A E Van Tu D Sri-Aroon P Allen D 2012 The status and distribution of freshwater molluscs of the Indo-Burma region. In The Status and Distribution of Freshwater Biodiversity in Indo-Burma Allen D Smith K Darwall W *Eds Gland Cambridge UK* 66–89 pp.

[13] Coates D Poeu O Suntornratana U Tung N T Viravong S 2003 Biodiversity and Fisheries in the Mekong River Basin *IWA Publishing London UK.*

[14] Aldea C Olabarria C Troncoso J S 2009 Spatial patterns of benthic diversity in molluscs from West Antarctica. *Antar Scie* **21**(4) 341-353.

[15] Albuquerque F Peso-Aguiar M Assunção-Albuquerque M Gálvez L **2009** Do climate variables and human density a_ect *Achatina fulica* (Bowditch) (Gastropoda: Pulmonata) shell length, total weight and condition factor? *Braz J Biol* **69** 879–885. [CrossRef]

[16] Sor R Ngor P B Boets P Goethals P L M Lek S Hogan Z S and Park Y S 2020 Patterns of Mekong Mollusc Biodiversity: Identification of Emerging Threats and Importance to Management and Livelihoods in a Region of Globally *Sign Biodi and Endem.*

[17] Alfred J B Varshney R K and Ghosh AK 1997 An assessment manual for faunal biodiversity sin south Asia. *SACEP/NORAD public seri on Biodiv in South Asia.*

[18] Omori M and Ikeda T 1984 Methods in marine zooplanktonecology. *John Wiley and Sons Inc New York* 256 pp.

[19] Serafim J M Lansac-Tôha F A Paggi J C Velho L F M and Robertson B 2003 Cladocera fauna composition in a river-lagoon system of the upper Paraná River floodplain, with a newrecord for Brazil. *Brazilian Journal of Biology* **63**(2) 349-356.

[20] Floder S and Sommer U 1999 Diversity in plankton communities: An experimental test of the intermediate disturbance hypothesis. *Limnol and Oceano*, 44(4): 1114-1119.

[21] PrtoNeto,V.F.2003.Zooplankton as bioindicator of environmental quality In the Tamandane Reef system (Pernambuco - Brazil): Anthropogenic influences and interaction with mangroves .Ph.D. thesis , University of Bremen,Brazil.

[22] Magurran, A. E. 2004. Measuring Biological Diversity. BlackwellPublishing, Oxford, 256 pp.

[23] Neves,I.F,Rocha,D.,Roche,K.F. and Pointe ,A.A. (2003).Zooplankton community structure of two mariginal lakes of river (Cuiaba)(Mato.Grosso,Brazil) with analysis of rotifera and cladocera diversity *Braz Biol* **63**(2) 329-343.

[24] Al-Kinani D M A H and Al-Mukhtar, I A A H 2014 Evaluated the biodiversity guide for the benthic invertebrates community and its relationship with the variability of some environmental factors in the Tigris and Diyala River within Baghdad City *Bagh J Scie* **2** (11).

[25] Faris A M 2010 The abundance of various tentacles and some animal wanderers and their diversity in the northern part of the Shatt al-Arab *Mas Thesis, College of Education / Basrah University* 119 p.

[26] Khalaf R Z 2011 Ecological Study of Gastropods from Intertidal Zone of Shatt Al-Arab / Iraq. *Msc thesis College of Science University of Basrah* 94 p.

[27] Al-Khafaji K Kh S Al-Essa S A K and Hashem A A 2015 The Diversity, Abundance and Distribution of Macroinvertebrates community in the intertidal zone of the Shatt-Al-Arab, South-Iraq *J Scie of Dhi Qar* **50** (2).

[28] Rabee A M 2007 Biodiversity of Rotifera and Cladocera in the upper region of Euphrates River- Iraq. *Umm Salamah J Scie* **4** (2): 221 - 232.

[29] Rabee A M and Al-Abayji J K 2009 Use of biodiversity indicators for water invertebrates to assess the health of the water in the Tigris andEuphrates rivers, central Iraq *Third scientific conference of College of Science University of Baghdad*

[30] Mirza N N A Nashaat M R 2019 Abundance Diversity and Distribution of Mollusca in the Gharaf River Southern Iraq . *Iraq J Scie* **60** (3):469-485 Pp.

[31] Al-Zurfi S K Suhad H H and Anam A T 2019 Biodiversity Study of Zooplanton in Selected Bahr Al-najaf Depression Najaf Governorate, Iraq *Bull Iraq nat Hist Mus J* **15** (3):263-278

[32] Al-Maliki N S 2002 Environmental survey of the Hamdan Canal, one of the main branches of the Shatt al-Arab River. *Msc Thesis College of Agriculture Basrah University.* 71p.

[33] Akbar M M 2013 Ecological Survey of Al-Gharaf Canal at Thi Qar Province, Iraq. *Marsh Bull* **8**(1) 1-17 .

Study of insect diversity in areas surrounding The Central Marsh in southern Iraq

Ali Kareem[1], Raad K. Aljaafari[1] ,Manar A. Abaas[1], Sienaa M. Al-Zurfi[1] ,Nadheer A. Fazaa[2,4], Omar F. Al-Sheikhly[3], and Mukhtar K. Haba [2] Mark J. Whittingham [4]

[1] Agriculture College, Kerbala University, Iraq.

[2]College of Sciences for Women, Baghdad University, Iraq.

[3]College of Sciences, Baghdad University, Iraq.

[4]School of Natural and Environmental Sciences, Newcastle University, UK

Corresponding Author: **ali.kareem@uokerbala.edu.iq**

Abstract

During the several visits in 2018 and 2019 to some areas surrounding Central Marsh in southern Iraq (CMSI) for conservation purposes. The fauna of insects from different sites and sessions of the CMSI are surveyed in this study; there were (18) species, (17) genera that belong to (16) families and (7) orders which are recorded in this investigation using sweeping nets and light traps. According to available references, 7 species are new records in the area of study. This is survey is important for updating insects diversity in central Marshlands in Iraq. A counting study is needed to understand more about insect distributions and their impacts on biodiversity in such an important area.

Keywords: Insects, Diversity, The Central Marsh, Fauna, Iraq.

1. Introduction

The Mesopotamian Marshes in the south of Iraq consist of three main marshes: Central Marsh, Al-Hawiza Marsh, and Al-Hammar Marsh (10,000 km2 during the 1970s) [1]. The Marshlands are considered one of the biggest and ancient wetlands in the Middle East, world heritage sites, and very important key biodiversity areas (KBAs) in Iraq [2]. The natural systems of the Mesopotamian Marshlands have faced significant challenges of water level during the last 40 years ago [3].

Biodiversity is the variety of species in a specific area [4]. The level of species diversity in the Mesopotamia marshes was observed within the variety of fluctuation among the flood and drainage seasons [5, 6]. The macroinvertebrates including insects are good indicators of current human-driven climate change and ecosystem [7, 1]. They have responded to global warming in all phenology and distribution to undergoing evolutionary changes [3].

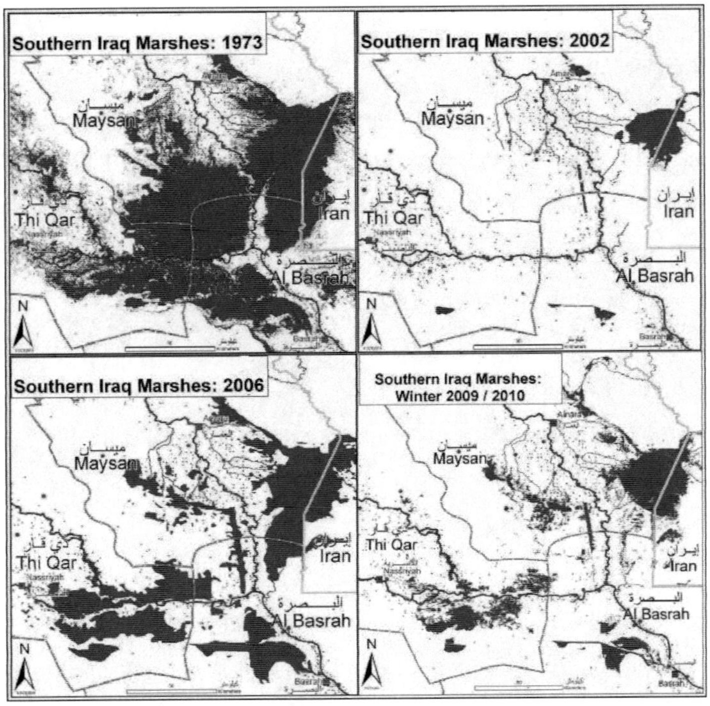

Figure 1. The water level in Mesopotamia marshlands over 40 years [1] (IMRP, 2006).

There are many studies carried out to survey insects diversity in Eastern Al-Hammar Marsh. Twenty-two species of Coleoptera were reported from [8], Four species of Hemiptera [3], Many species of Odonata [9], and about 10 species of Diptera [3,10,9] have been recorded. According to available data, there is no specific study to show the diversity of insects in the CMSI. Therefore, it is needed to know insects species diversity in this important area in Iraq as a part of its conservation which is the aim of this study.

2. Materials and Methods

This study was carried out during 2018-2019 in the area surrounding the Central Marshland in Iraq. The specimens of insects (adults only) were collected from different sites from the Central Marsh in Southern Iraq (CMSI) using the sweeping net and light traps; the whole samples studied were labelled and stored in 70% ethanol and kept in the Entomology laboratory at the Faculty of Agriculture, Kerbala University.

The collected specimens were sent to Iraq Natural History Museum and Research Center (NHMRC), the University of Baghdad to identify and confirm. The high taxa identified using available keys for Order and Families. The lowest taxonomic was confirmed by Drs Razzaq Augul and Hanaa Al-Saffar in the NHMRC.

3. Results

In the present study, there were (18) species from (17) genera belonging to (16) families and (7) orders, which were sampled from different locations and sessions in CMSI. According to available references, 7 species are new records in the area of study.

3.1 Order 1: Coleoptera

Family1: Coccinellidae,

Coccinella septempunctata (Linnaeus, 1758), (Seven-spot Ladybird)

Specimens examined: Al-Chabiish (Nasiriyah) 2018 (2 ♂ + 2 ♀) Coordinated 31°03′N 46°16′E.

Global distribution: Widespread species in Europe, Asia and Northern Africa [11], including Iraq [12].

Family2: Hydrophilidae,

Hydrophilus mesoptamia Knish,1914

H . nubilus F. Kuweit. 1886 (= *H. mesopotanziae* Kuw.), :Euphrates ; Knisch, 1924 [13].

Specimens examined: Al-Chabiish (Nasiriyah) 2018 (2 ♂ + 2 ♀) Coordinated 31°03′N 46°16′E.

Global distribution: Widespread species and in the middle in the south in Iraq [13].

Family3: Dynastidae,

Pentodon algerinus Füssly, 1778

Specimens examined: Al-Chabiish (Nasiriyah) 2018 (2 ♂ + 2 ♀) Coordinated 31°03′N 46°16′E.

Global distribution: Portugal and North Africa to northwestern China [14], UAE [15]. According to available references, this species **new record** in Al-Chabiish (Nasiriyah) Iraq (this study).

Family4: Dytiscidae,

Cybister tripunctatus Oliver, 1795

Specimens examined: Al-Chabiish (Nasiriyah) 2018 (2 ♂ + 2 ♀) Coordinated 31°03′N 46°16′E.

Global distribution: Widespread species in Europe, Asia and Croatia [16]. According to available references, this species **new record** in Al-Chabiish (Nasiriyah) Iraq (this study).

Family 5: Tenebrionidae,

Subfamily 1: Tenebrioninae LATREILLE, 1802

Tribe: Blaptini LEACH, 1815

Genus: Blaps Fabricius, 1775

Blaps (*Blaps*) *gigas* (LINNAEUS, 1767)

Distribution: Southern and Central Europe, Turkey, Iraq, Kazakhstan.

Specimens examined: Al-Chabiish (Nasiriyah) 2018 (2 ♂ + 2 ♀) Coordinated 31°03′N 46°16′E.

Global distribution: Southern and Central Europe [11], Turkey, Iraq, Kazakhstan [17].

Subfamily 2: Pimeliinae

Tribe: Adesmiini

Adesmia cancellata (*clathrate*) Solier, 1835

Specimens examined: Al-Chabiish (Nasiriyah) 2019 (2 ♂ + 2 ♀) Coordinated 31°03′N 46°16′E.

Global distribution: Asia: Iran, Iraq, Oman, Saudi Arabia, United Arab Emirates [18], Egypt [19].

IOP Conf. Series: Earth and Environmental Science 1002 (2022) 012002 doi:10.1088/1755-1315/1002/1/012002

Family 6: Carabidae, (Granuad beetls)

Calosoma olivieri Dejean, 1831

Specimens examined: Al-Chabiish (Nasiriyah) 2018 (2 ♂ + 2 ♀) Coordinated 31°03′N 46°16′E.

Global distribution: AF, DZ, EG, IQ, IR, IT, JO, LY, MA,

PK, SA, SY, TD, TM, TN, UZ, YE. PAL_SAR species [20].

Carabidae,

Scarites eurytus Fisher von Waldheim,1828

Specimens examined: Al-Chabiish (Nasiriyah) 2018 (2 ♂ + 2 ♀) Coordinated 31°03′N 46°16′E.

Global distribution: Widespread, Iran, Iraq [21].

Family 7: Buprestidae,

Capnodis miliaris Klug, 1829

Specimens examined: Al-Chabiish (Nasiriyah) 2018 (2 ♂ + 2 ♀) Coordinated 31°03′N 46°16′E.

Global distribution: Jordan, East Mediterranean-Turanian, Iraq, Cyprus, Southern, Italy, Lebanon, Greece Palestine, Syria, Turkey, Iran and USSR. [22, 23].

.

3.2 Order 2: Orthoptera

Family1: Acrididae,

Aiolopus strepens (Latreille, 1804)

Acrididae, *Aiolopus simulatrix* (Walker, 1870)

Acrididae, *Anadridium aegyptium* (Linnaeus, 1764).

Acrididae, *Tropidopola cylindrica* (Marschall, 1836).

Specimens examined: Al-Chabiish (Nasiriyah) 2018 (2 ♂ + 2 ♀) Coordinated 31°03′N 46°16′E.

Global distribution: Widespread in the Mediterranean and reported from southern Europe and Asia and Iraq [24].

Family 2: Gryllidae,

Acheta sp. Linnaeus, 1758

Specimens examined: Al-Chabiish (Nasiriyah) 2018 (2 ♂ + 2 ♀) Coordinated 31°03′N 46°16′E.

Global distribution: Widespread species including Iraq [25].

Family 3: Grullotalpidae,

Grullotapa gryllotalpa (Linnaeus, 1758).

Specimens examined: Al-Chabiish (Nasiriyah) 2018 (2 ♂ + 2 ♀) Coordinated 31°03′N 46°16′E.

Global distribution: Widespread species including Iraq [25].

3.3 Order 3: Odonata

Dragonfly and Damselflies

Family1: Lestidae,

Sympecma paedisca (Brauer, 1877)

The Siberian Winter Damselfly

Specimens examined: Al-Chabiish (Nasiriyah) 2018 (2 ♂ + 2 ♀) Coordinated 31°03′N 46°16′E.

Global distribution: Western Europe, Asia, France, Germany, Italy [26]. According to available references, this species **new record** in Al-Chabiish (Nasiriyah) Iraq (this study).

Family2: Aeshnidae,

Anax ephippiger (Burmeister, 1839)

Specimens examined: Al-Chabiish (Nasiriyah) 2018 (2 ♂ + 2 ♀) Coordinated 31°03′N 46°16′E.

Global distribution: It is a migrant insect. Ranging from Africa and South-West Asia to Europe, Central and East Asia, reaching Iceland in the north, China, Japan and Thailand in the east and crossing the Atlantic Ocean to reach the Caribbean region and even Brazil [27]. According to available references, this species **new record** in Al-Chabiish (Nasiriyah) Iraq (this study).

3.4 Order 4: Mantodea

Family: Empusidae ,

Blepharopsis mendica Fabricius, (1775)

Specimens examined: Al-Chabiish (Nasiriyah) 2018 (2 ♂ + 2 ♀) Coordinated 31°03′N 46°16′E.

Global distribution: Recorded in Iraq but not in Al-Chabiish (Nasiriyah) according to available references, Asia Minor, Iran, Cyprus, Northern Africa and Canary Islands [28].

3.5 Order 5: Hymenoptera

Family: Eumenidae,

Polistes wattii Cameron, 1900

Specimens examined: Al-Chabiish (Nasiriyah) 2018 (2 ♂ + 2 ♀) Coordinated 31°03′N 46°16′E.

Global distribution: Southwest, Iran, India, and South China, and Arabian Peninsula [29]. According to available references, this species **new record** in Al-Chabiish (Nasiriyah) Iraq (this study).

3.6 Order 6: Lepidoptera

Family: Nymphalidae,

Vanessa cardui (Linnaeus, 1758).

Specimens examined: Al-Chabiish (Nasiriyah) 2018 (2 ♂ + 2 ♀) Coordinated 31°03′N 46°16′E.

Global distribution: This is a very strong migrant species as well as a very strong migrant found in North and south Africa, Australia, Mt Imittos and Venezuela [30]. According to available references, this species **new record** in Al-Chabiish (Nasiriyah) Iraq (this study).

3.7 Order 7: Diptera

Family: Ephydridae

Scatella tenuicosta Collin, 1930

Specimens examined: Al-Chabiish (Nasiriyah) 2018 (2 ♂ + 2 ♀) Coordinated 31°03′N 46°16′E.

Global distribution: Russia, Europe [31]. According to available references, this species **new record** in Al-Chabiish (Nasiriyah) Iraq (this study).

4. Discussion

The Central Marsh in southern Iraq (CMSI) was one of the very important conservation areas in Iraq. In the protected areas, biodiversity monitoring is vital for active management. The insect's fauna of these protected areas has involved relatively little attention as compared to those of invertebrates [32], although international conventions signed by Iraq as a member of the Convention on Biological Diversity since 2001[33]. Many species of insect reported as new for Iraq fauna [34, 35, 36, 37, 38, 39]. Eighty species of 17 genera that belong to 16 families and 7 orders of insects were recorded in this study. According to available references, 7 species are new records in the area of study. All recorded species of insects from the CMSI are different from those recorded in Eastern Al-Hammar Marsh by [10, 9, 3]. This study is primarily the result of insect diversity in CMSI. Beyond enhancing our knowledge of CMSI insects fauna, these results will provide useful information for guiding the conservation activities [40, 41] in the CMSI and starting point to the future more detailed investigation on the fauna in Iraq marshes.

Acknowledgements

The authors would like to thanks Green Climate Organization (GCO) and Al-Chabiish Organization for Ecology Tourists for coordinating the trips to Marshlands. Also, we would like to thanks Drs Razzaq Augul and Hanaa Al-Saffar from the NHMRC for helping to confirm the species.

References

[1] IMRP (Iraq Marshlands restoration Programme) (2006). Monitor marsh Ecosystem Recovery. Final report. USAID/development Alternative Inc. 528pp.

[2] Kohring, M., 2019. Key Biodiversity Areas of Iraq: Priority Sites for Conservation and Protection Nature Iraq. Natural Areas Journal, 39(2), pp.275-276.

[3] Hussain N.A. (2014). Biotopes of the Iraqi marshes. First Edition. Dhifaf publishing house, Basra, Iraq. 432pp.

[4] Br nmark, C., & Hansson, L.-A. (2005). The biology of lakes and ponds. Oxford: Oxford University Press.

[5] Al-Hilli M.R. (1977). Botanical studies of the marshes of southern Iraq. PhD. Thesis. University of Cairo, Egypt. 460pp. (in Arabic).

[6] Richardson CJ ReissP Hussain NA Alwash AJ Pool DJ. 2005. The restoration potential of the Mesopotamian marshes of Iraq. Science. 307: 1307– 1311.

[7] ARDI (Agriculture Reconstruction and Development Programme of Iraq) (2006). Monitor Marsh Ecosystem recovery, Final Report. USAID/Development Alternative, Final report 177pp.

[8] Abudal-Karim R.M. (1987). An introduction to the taxonomy of the family Dytiscidae (Coleoptera) of Iraq. M.Sc. thesis, Basrah University. 152pp.

[9] Geraci, C.J., Zhou, X. and Al-Saffar, M. 2011. Barcoding Iraq: Aquatic Insects of the Tigris/Euphrates River Basin Useful for Biosurveillance. 4th International Barcode of Life Conference, Adelaide, Australia.

[10] Hassan K.S., Habeeb M.A., Al-Mousawi N.J. (2001). Occurrence of aquatic insects with algae in Basrah province. Marine Mesopotamica 15(1): 137-143.

[11] CABI (2021).*Coccinella septempunctata*. Retrieved from https://www.cabi.org/isc/datasheet/11733#REF-DDB-125778 (Accessed: 02.2021).

[12] Amin, A.H. and Muhammed, S.H., 2008. Seasonal abundance of mealy plum aphid, *Hyalopterus pruni* (Geoffroy) and its natural enemies on some stone fruit trees in Erbil City, Kurdistan Region, Iraq. Egyptian Journal of Biological Pest Control, 18(1), pp.249-256.

[13] Abdul-Rassoul, M.S. and Ali, H.A., 1999. Hydrophilidae of Iraq (insect, Coleopter). Bulletin of the Iraq Natural History Museum, 9(1), pp.1-5.

[14] Krell, F.-T. (2006): Dynastinae. Pp. 277–283 in I. Löbl & A. Smetana (eds), Catalogue of Palaearctic Coleoptera. Vol. 3. Stenstrup: Apollo Books.

[15] Krell, F.T. and Král, D., (2017). Order Coleoptera, family Scarabaeidae. Arthropod fauna of the UAE, 6: 169–185.

[16] Temunović, M. and Šerić Jelaska, L., 2009. First record of diving beetle *Cybister tripunctatus africanus* Laporte, 1835 in the Croatian Hydradephagan fauna. Communications and abstracts, p.85.

[17] Bunalski, M. and Ghahari, H., 2017. Contribution to the study of darkling beetles (Coleoptera: Tenebrionidae) from Golestan Province, Northern Iran. Wiadomości Entomologiczne, 36(1), pp.25-31.

[18] BioLib (2021) Available at: https://www.biolib.cz/en/taxon/id860928/ (Accessed at: 03.2021).

[19] Elshewy, D., Salem, M. and Elmetwally, N.E. (2016) Checklist of the family Tenebrionidae (Coleoptera) in Egypt. Egypt. J. Agric. Res., 94 (1).pp 39-57.

[20] Abdel-Dayem, M.S., Elgharbawy, A.A., Rasool, I., Nagel, P. and Aldhafer, H.M. (2019) 'The Carabidae (Coleoptera) of Shada Al-A'Ala Nature Reserve, Southwestern Saudi Arabia, with description of a new species of Paussinae', J ZooKeys, (812), p. 93.

[21] Ali, H.A. (1964) *An introduction to the taxonomy of Iraqi Carabidae Col., with an examination of the taxonomic value of internal characters*. Imperial College of Science and Technology.

[22] Al-Hariri, G. 1972. Economic Insects in Syria and the neighbouring countries (In Arabic). University of Aleppo, Syria. 465 pp.

[23] Curletti, G. (1994) 'I Buprestidi d'Italia: catalogo tassonomico, sinonimico, biologico, geonemico', *Natura Bresciana*, 19, pp. 1-318.

[24] COPR (Center for Overseas Pest and Research). 1982. The locust and Grasshopper Agricultural Manual. Centre for Overseas Pest and Research; London. 690 p.

[25] Khalaf, K. (1959) 'A collection of insects from Iraq', *J The Iraq Natural History Museum Publication*, (17), pp. 17-26.

[26] Battisti, A. and Pavesi, M. (2017) 'First records of breeding *Sympecma paedisca* (Brauer, 1877)(Odonata Lestidae) in Italy', *J Biodiversity Journal*, 8(2), pp. 763-768.

[27] Kosterin, O. and Borisov, S. (2018) 'New records and migration strategy of *Anax ephippiger* (Burmeister, 1839)(Odonata, Aeshnidae) in the territory of the Russian Federation', *J Евразиатский энтомологический журнал*, 17(1), pp. 73-79.

[28] Mohammad, S.K., Alla, S.M.G., El-Hamouly, H., Ehrmann, R. and EL-DEN NASSER, M.G. (2011) 'Mantodea of Egypt', *J Zootaxa* 3044(1), pp. 1-27.

[29] GBIF, 2021 .*Polistes wattii* Cameron, 1900. https://www.gbif.org/species/1310601 (Accessed at: May 2021).

[30] Alexiou, S. (2014) 'The Bυττεrflies (Lepidoptera: Papilionoidea & Hesperioidea) of MT. Imittos, Attiki, Greece', *Parnassiana Archives*, 2, pp. 25-52.

[31] MacGowan, I., Vikhrev, N., Krivosheina, M., Ruchin, A. and Esin, M. (2021) 'New records of Diptera from the Republic of Mordovia, Russia', *J Far Eastern Entomologist*, 9(423).

[32] Abuzinada, A.H., Al-Wetaid, Y.I. and Al-Basyouni, S.Z.M., 2005. Convention on Biological Diversity. The National Strategy for Conservation of Biodiversity in the Kingdom of Saudi Arabia. The National Commission for Wildlife Conservation and Development, Riyadh, Saudi Arabia, 94.

[33] CBD, Convention on Biological Diversity 2001. 5th National Report to the Convention.

[34] Kareem, Ali A., Manar A. Abbas, Raad K. Aljaafari, Sienaa M. Al-Zurfi, and Zeina M. Mouhsan. 2019. "New report of Hoverfly parasite *Diplazon laetatorius* (Fabricius, 1781)(Ichneumonidae-Hymenoptera) from Iraq." In IOP Conference Series: Earth and Environmental Science, vol. 388, no. 1, p. 012001. IOP Publishing.

[35] Polaszek, A., Kareem, A.A., Aljaafari, R.K., Al-Zurfi, S.M. and Abbas, M.A., 2019. New records of *Scelio* species (Hymenoptera: Scelionidae) from Iraq: egg parasitoids of acridid grasshoppers (Orthoptera: Acrididae). In IOP Conference Series: Earth and Environmental Science (Vol. 388, No. 1, p. 012004). IOP Publishing.

[36] Al-Zurfi, S.M., Aljaafari, R.K., Abbas, A. and Kareem, A.A., 2020. Reporting *Synechocryptus bovei* (Brullé, 1846)(Hymenoptera: Ichneumonidae: Cryptinae) first time in Iraq. In IOP Conference Series: Earth and Environmental Science (Vol. 553, No. 1, p. 012024). IOP Publishing.

[37] Kareem, A.A., Lahuf, A.A., Aljaafari, R.K. and Buhl, P.N., .2020. First report of the parasitoid wasp *Platygaster oebalus* (Hymenoptera: Platygastridae) from Iraq. In IOP Conference Series: Earth and Environmental Science (Vol. 553, No. 1, p. 012004). IOP Publishing.

[38] Kareem, A., Raad Kareem Aljaafari, Sienaa Muslim Al-Zurfi, Muntather M. Almosawy and Zeina M. Mouhsan (2020a) 'New record of *Dichrogaster modesta* (Gravènhorst, 1829) (Hymenoptera: Ichneumonidae: Cryptinae) in Iraq', *Ecology, Environment and Conservation*, 26(1), pp. 75–77.

[39] Kareem, A.A., Lotfalizadeh, H. and Aljaafari, R.K. (2020b) 'First record of *Leucospis bifasciata* Klug (Hymenoptera: Leucospidae) in Iraq', *Journal of Insect Biodiversity Systematics*, 6(3), pp. 223-228.

[40] Koivula, M.J. (2011) 'Useful model organisms, indicators, or both? Ground beetles (Coleoptera, Carabidae) reflecting environmental conditions', *J ZooKeys*, (100), p. 287.

[41] Kotze, D.J., Brandmayr, P., Casale, A., Dauffy-Richard, E., Dekoninck, W., Koivula, M.J., Lövei, G.L., Mossakowski, D., Noordijk, J. and Paarmann, W. (2011) 'Forty years of carabid beetle research in Europe–from

taxonomy, biology, ecology and population studies to bioindication, habitat assessment and conservation', *J ZooKeys*, (100), p. 55.

Effect of Enzyme Supplementation in Diet on Growth of Common Carp (*Cyprinus carpio* L.) juveniles in RAS

Noori Abdul-Nabi Nasir[1], AbdulKareem Taher Yesser[2] & Qusay Hamid Al-Hamadany[3]

Department of Vertebrates, Marine Science Center, Basrah University, Basrah, Iraq.
* noori.nasir@uobasrah.edu.iq
00964 7713167345
*https://orcid.org/0000-0003-1667-2752

** https://orcid.org/0000-0002-6963-4082

***https://orcid.org/0000-0002-3963-9472

Abstract
The purpose of this study was to determine the effect of supplementing the diet with Thepax Aqua (Aqualase) on the development and survival of juvenile common carp (*Cyprinus carpio* L.). Therefore, an eight-week nutrition experiment was carried out using the Thepax Aqura enzyme in the Department of Vertebrates, Marine Science Center, University of Basrah. This enzyme was supplemented at 1, 2, and 3% in the control diet. The maximum weight growth (3.9870 gm 0.1230) and Specific Growth Rate (0.4238 gm 0.0049) were seen in the fish fed with T1 diet, and these results were found to be statistically different ($P<0.05$) from those observed in the fish fed with T2 (2 %), T3 (3 %), and T0 (control) diets. This study also concluded that there was no significant effect ($P\leq 0.05$) for enzyme levels on the survival of juveniles carp fed with different dietary enzymes. Finally, the outcome of the current experiment showed that the 1% supplementation of Thepax Aqua to the food gave good growth and feed utilization of Juveniles of carp *C.carpio*.

Keywords: Growth, survival, juveniles, *Cyprinus carpio*, enzyme, RAS.

1. Introduction

The study was conducted to determine the influence of Thepax Aqua Common Carp (*Cyprinus carpio*) farming becoming a vital sector to meet the increasing demand for carp worldwide [1]. The feed of the fish is the primary factor in fish culture, which reached about 60% of the total production cost [2]. Nutrition can influence several mechanisms controlling the growth of the fish juveniles. However, the growth development of fish juveniles can be influenced by a number of genetic and ecological factors, particularly the occurrence and content of feeds. Digestion effectiveness of the farmed fish could be improved by using the enzymes to their food [3]. Additionally, various scientists have demonstrated that supplementing food with the enzyme significantly improves the development performance and feed consumption of juvenile fish [4, 5]. Later, Klein *et al.* [6]. reported that the digestion of nutrients is influenced by the digestive enzymes' action, which is accountable for the final phases of breaking down and absorbing the food. On the other hand [3]. stated that the digestion effectiveness of the cultivated species can be better by adding the enzymes to the feed. There are few details on the use of enzymes in the food of Juveniles of *C. carpio*. There was no information on the use of Thepax Aqua in the diet for juveniles of *C. carpio*. As a result, the current investigation was designed to determine the role of the dietary Thepax Aqua enzyme in the growth and survival of juvenile *C. carpio*.

2. Materials and Methods

120 Juveniles of *C. carpio* with an average weight ranging from 12.817gm ±0.647 to 13.912 gm ± 1.1756 were taken from the Marine Science center fish ponds, the University of Basrah the aquaculture unit of Marine Science Center. The experiment was carried out using the pilot Recirculation Aquaculture system (RAS) with 12 rearing glass aquaria (40×70×60 cm) for 60 days. Mechanical filtration was installed in this system. During the experiment, the dissolved oxygen was kept at 6 - 8 mg/L, the pH was kept at 7.3- 7.6, and the temperature was kept at 25 -28 °C.The fish were fed twice a day (9.00 a.m. and 2.00 p.m.) at a rate of 3% of their body weight with a control diet and the other three diets with varying levels of enzymes. The weight of the experimental fish was measured at the beginning and end of the experiment, as well as every two weeks during the experiment. Fences and uneaten feed were removed from the tank daily at 8 a.m. by siphoning out 25% of the water and replacing it with freshwater. During the experiment, aeration was also provided to protect the fish from the stress condition by using one diffuser air stone for each tank [2].

2.1. Diet preparation

Control diet

Control diet was prepared by use Fishmeal, Soya meal, Wheat flour, Corn, Vitamins, Lysin and Methonin [7]. The components of the control and test diets are listed in Table 1.

Table 1. The composition (%) of experimental diet used for the *Cyprinus carpio* during 60 days period.

Ingredients %	Diet 1 T1	Diet 2 T2	Diet 3 T3	Control TO
Fish meal	32	32	32	32
Soya meal	22	22	22	22
Wheat flour	29	29	29	29
Corn	14	13	12	15
Vt.& Minerals	1	1	1	1
Lys.	0.5	0.5	0.5	0.5
Meth.	0.5	0.5	0.5	0.5
Enzyme *	1	2	3	0
Total	100	100	100	100

* THEPAX Aqua (AQUALASE) , each 1 kg contains (100% inactivated Saccharomyces cerevisiae. Var Ellipsoid strain. With count 50 * 109 / gm. 16% beta glucan. 18% Manna oligosaccharides).

2.2. Experimental diet

The enzyme Thepax Aqua (Italian DOX LTD.) was dissolved in 60 ml distilled water and mixed cautiously in the food and then supplemented at the rate of 1, 2 and 3 % in control diet and used as test diet which presented as T1, T2 and T3 and its value was zero for the control group (Table 1).

2.3. Ingredients and approximate composition

Table 2 depicts the approximate compositions of the control and test diets. Moisture, Protein, Fat, Ash, Carbohydrate, and Gross energy were measured in each diet [8].

Table 2 shows the approximate composition (percent dry weight basis) of the diets fed to juvenile *Cyprinus carpio* for a period of 60 days.

Components	Diet 1 T1	Diet 2 T2	Diet 3 T3	Control T0
Moisture %	9.7	9.95	10.19	9.7
Protein %	29.1	28.82	28.38	29.01
Fat %	7.91	7.68	7.54	7.91
Ash %	9.46	8.38	7.49	9.46
Carbohydrate %	43.83	45.27	46.4	43.92
Energy Mj/ kg	18.06	18.03	18.04	18.06

2.4. Growth parameters

The individual fish were weighed to nearest 0.01 mg at the starting and the finish of the experiment. At the conclusion of the experiment, the growth of the fish utilized in the feeding treatment was determined using the following formulas [1, 2 , 9]. as :

Weight gain (WG) = W2 (g) – W1(g)

Daily weight gain (DWG) = WG / t.

Relative growth rate (RGR%) = (W2 - W1/ W1) x 100

Specific growth rate (SGR %/day) = 100 (ln W2- ln W1) / t

Where W1: Average initial weight, W2: Average final body weights and t: Time of experiment (days).

Feed conversion ratio (FCR) = Food consumed (g) / WG (g).

Feed conversion efficiency (FCE) = WG (g)/ feed consumed (g) x 100.

Survival rate = Number of survive fish / number of fish at the beginning of experiment x100.

2.5. Economic parameters of the diets by calculating:

Profit index.

The profit index was calculated as:

Profit index (PI) = C_f (kg)/C_N, Cf and C_N are economic value of fish and diet, respectively [10].

Reduced cost (RC) = 100 - (100× (ECR diet / ECR Control diet), where ECR is economic coefficient ratio = FCR × Cost of feed [10].

2.6. Statistical Analysis

The results of the present work were analyzed using one-way ANOVA. Significant differences between groups were assessed at 0.05 level using IBM SPSS version 22 (Arlington, Virginia).

3. Results

Growth performance of juveniles *C. carpio* fed with the different enzyme diets is shown in Table 3. Weight gain, specific growth rate, Feed conversion rate and survival of juvenile C. *carpio* fed with control and experimental diets for 8 weeks are tabulated in Table 3. One-way ANOVA demonstrated significant difference (P<0.05) between the experimental diets for weight gain and specific growth rate.

Table 3. Growth parameters, Food conversion and Profit index of *Cyprinus carpio* juveniles after 60 days period.

Parameter	Treatment				Sign. *
	Diet 1	Diet 2	Diet 3	Control	
Initial Weight (g)	13.910 ±0.2711[a]	12.817± 0.647[a]	13.912±1.1756[a]	13.497 ±0.779[a]	-
Final weight (g)	18.870± 0.3900[a]	15.636± 1.0781[b]	16.662±1.1140[b]	16.057 ±0.7855[b]	*
Weight gain (g)	3.9870± 0.1230[a]	2.8183± 0.4391[b]	2.7493±0.2788[b]	2.7066 ±0.3558[b]	*
Daily weight gain (g/day)	0.0720± 0.0022[a]	0.0503± 0.0078[b]	0.0491±0.0049[b]	0.0484 ±0.0063[b]	*
Food conversion rate	2.1668± 0.2347[a]	2.2061±0.30045[a]	2.4309±0.2587[a]	2.5624 ±0.3001[a]	-
Food conversion efficiency (%)	46.1507± 2.3498[a]	39.0263± 2.7621[b]	41.1987±2.01905[bc]	45.3087±2.06400[ac]	*
Relative growth rate %	47.830± 0.6178[a]	39.1349±4.13492[b]	35.5399± 5.6322[b]	36.1444±3.8902[b]	*
Specific growth rate (g/day) %	0.4238± 0.00485[a]	0.3537± 0.0338[b]	0.3237± 0.0468[b]	0.3290 ±0.0325[b]	*
Reduced cost	15.6191± 1.2500[a]	14.1592± 0.1000[a]	8.4273± 1.0700[b]	0.000	*
Profit index	7.9235± 1.580[a]	7.7888± 1.6800[a]	7.3013± 1.4700[a]	6.686 ±1.8900[a]	-

*The mean difference is significance at the (P ≤ 0.05). Different litters in the same row show significance at P< 0.05.

As seen from Table3 and Figure1, the fish growth has significantly increased during the 8 weeks as the weight was about 13.910 g at the start of investigation and attained the weight of 18.870 g at the end of experiment by using 1% concentration of the enzyme and by no enzyme added (control diet), it was recorded to be 16.057g.

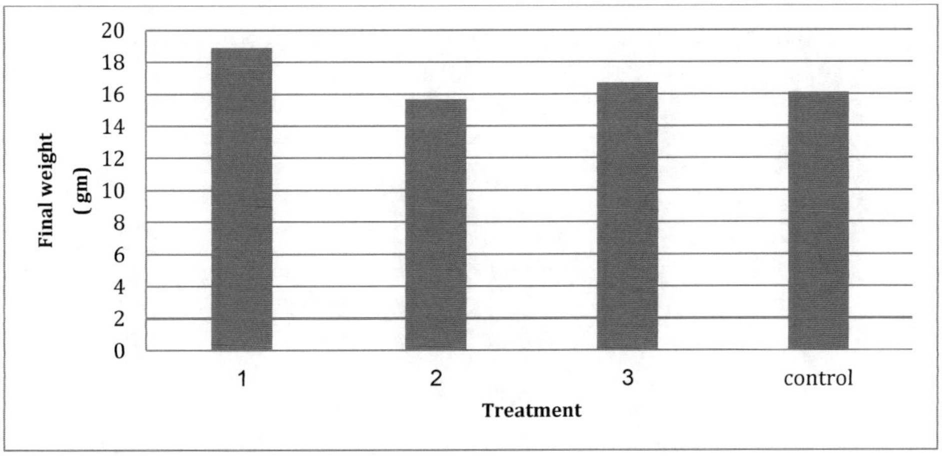

Figure 1. Changes in *Cyprinus carpio* juveniles weight after 8 weeks period.

Additionally, the results of this study suggested that fish fed with 1% Thepax enzyme had a significantly ($p < 0.05$) larger weight gain (3.987) when compared to fish fed with other treatments. (Figure 2). Therefore, the daily weight increase ($p<0.05$) was clearly shown in the diet containing 1 % enzyme (Figure 3).

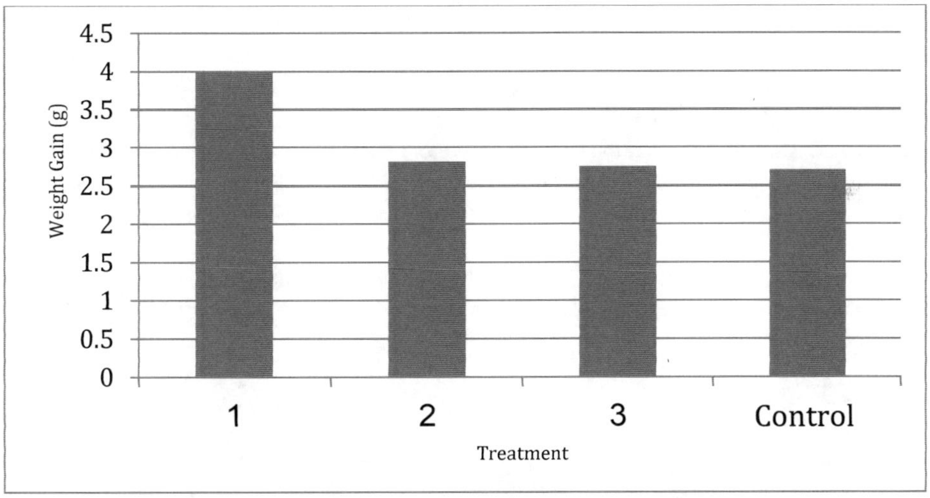

Figure 2. Weight gain for *Cyprinus carpio* juveniles after 8 weeks period.

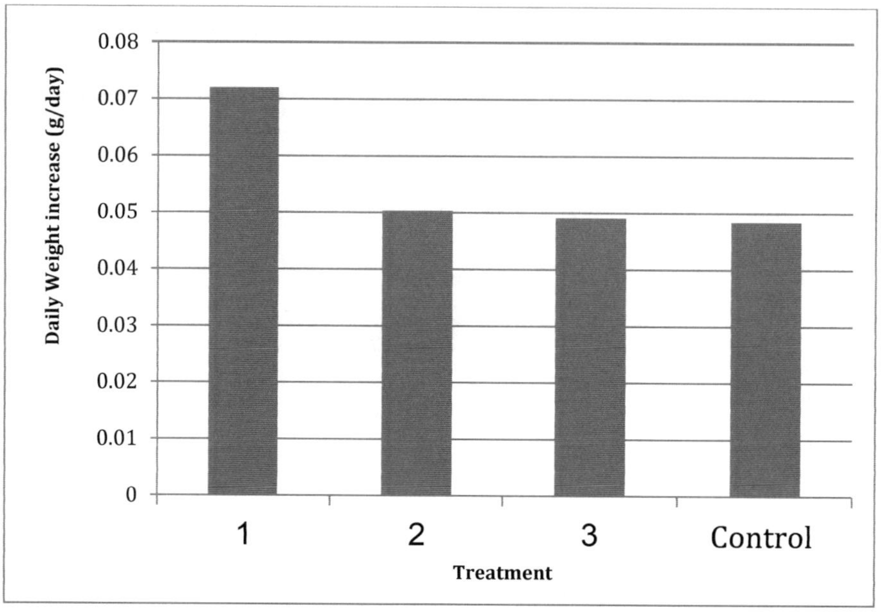

Figure 3. Daily weight increase weight for *Cyprinus carpio* juveniles after 8 weeks period.

The results also showed increase of Thepax enzyme supplementation in a diet more than 1 % results significant reduction in relative growth rate (Figure 4) and specific growth rate (Figure 5), therefore the larger concentration of Thepax enzyme offer no purpose in the diet.

The results were also indicated that the lower FCR (2.1668 ± 0.2347) values were observed in a diet containing 1% enzyme diet. The supplement of enzymes to diet had no significant effect on FCR of *C. carpio* juvenile. Furthermore; the values of FCR were increases among the 2% (T2), 3% (T3), 0.25% (control).Their values were 2.2061, 2.4309 and 2.5624 respectively (Figure 6).

It is obviously shown in Table 3 and Figure 7 that FCE increased ($p<0.05$) in fish fed by 1% enzyme diet and control diet but significantly decreased in 2 % and 3 % enzyme diet.

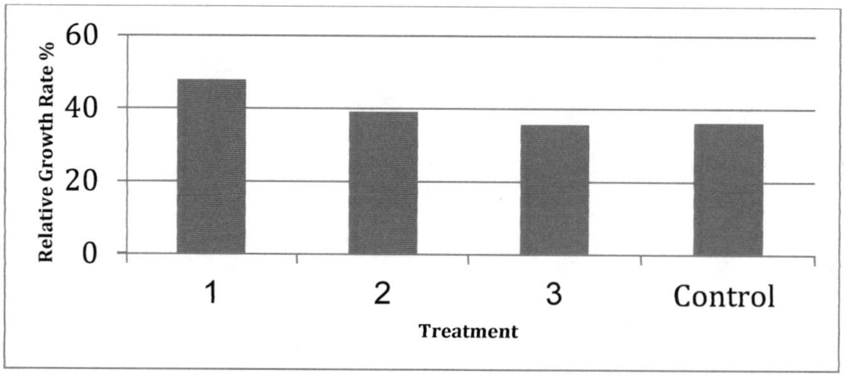

Figure 4. Relative growth rate (%) for *Cyprinus carpio* juveniles after 8 weeks period.

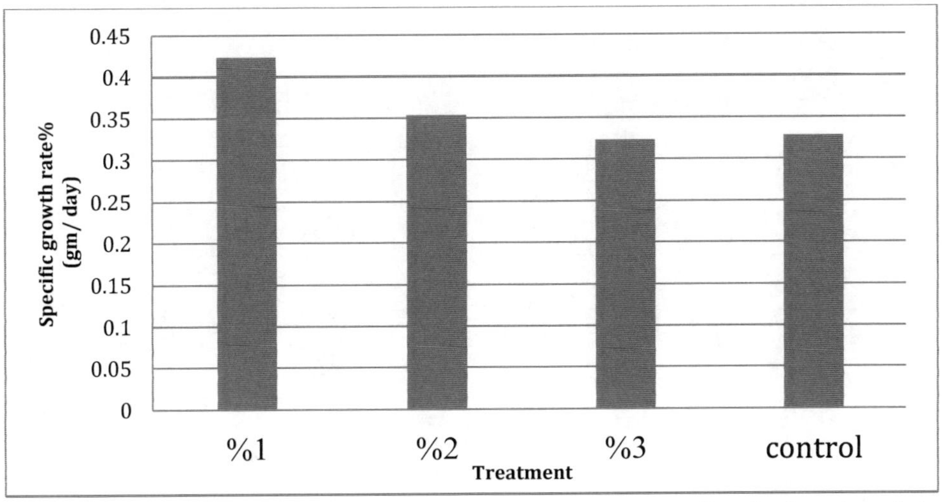

Figure 5. Specific growth rate (gm/ day) (%) for *Cyprinus carpio* juveniles after 8 weeks period.

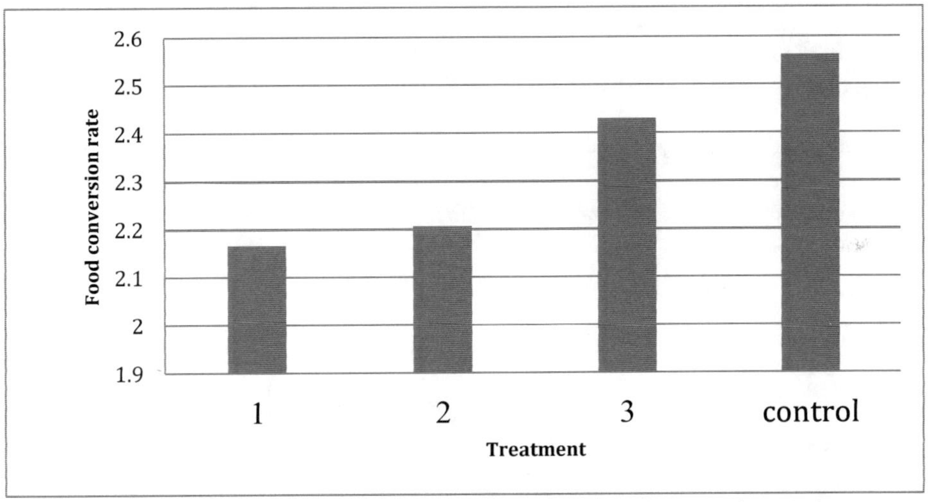

Figure 6. Food Conversion Ratio for Cyprinus carpio juveniles after 8 weeks period .

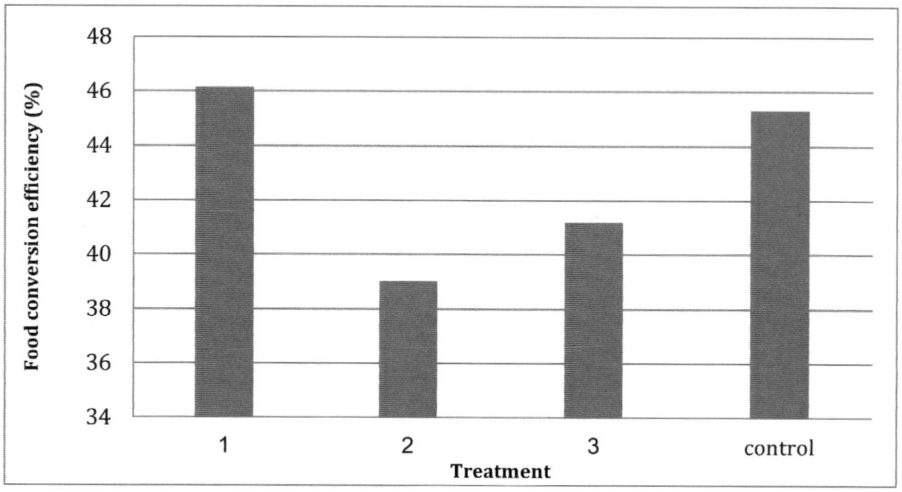

Figure 7. Food conversion efficiency (%) for *Cyprinus carpio* juveniles after 8 weeks period .

Cost-Benefit Analysis of Enzyme Supplementation in Diet

The use of dietary Thepax Aqua enzyme through the experiment led to reduce of feed cost (cost/kg feed) (Figure 8) and increase profit index by using diet 1 (Figure 9). However, the maximum cost reduction and maximum profit were achieved with diet containing 1% Thepax Aqua enzyme meal (Figure 8 and Figure 9).

The current study also found that maximum survival (57.17%) was seen in fish juveniles fed with 1% (T1) diet in 8 weeks investigation . The use of dietary Thepax Aqua enzyme during the experiment made survival rate high (100%) comparing with fish fed with diet control (80%) through 8 weeks rearing period as shown in Table (4).

Table 4. Survival Rate (%) for *Cyprinus carpio* juveniles after 8 weeks period.

Treatments	Periods (week)			
	2	4	6	8
Diet 1	100	100	100	90
Diet 2	100	100	100	100
Diet 3	100	100	100	100
Control	100	100	100	80

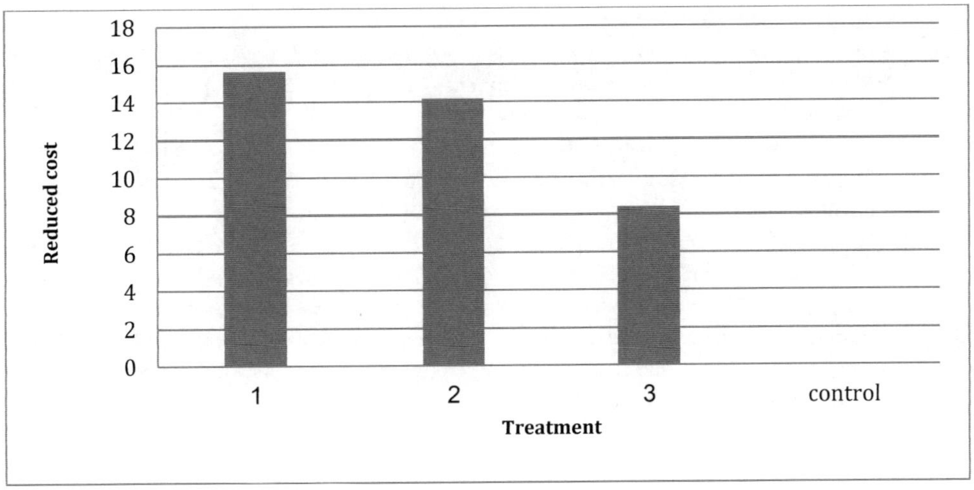

Figure 8. Reduced cost due to Enzyme Supplementation in Diet of *Cyprinus carpio* juveniles after 8 weeks period .

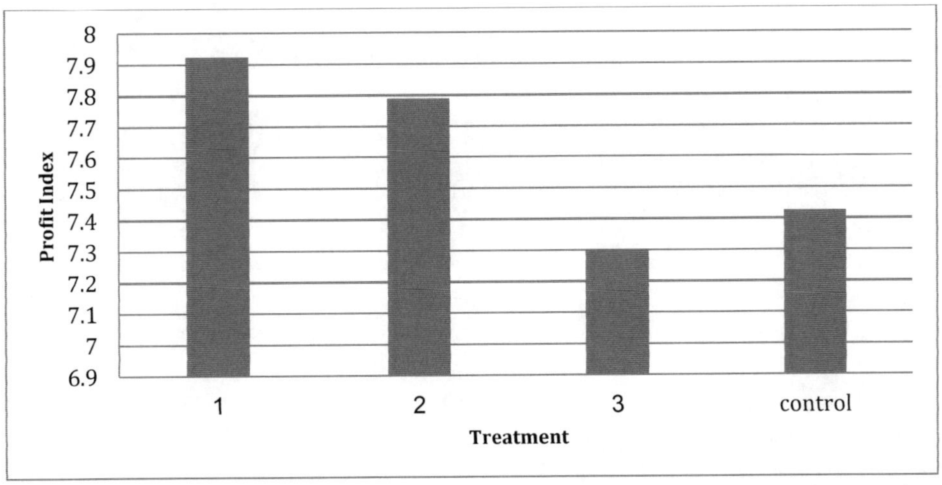

Figure 9. Profit Index due to Enzyme Supplementation in Diet of *Cyprinus carpio* juveniles after 8 weeks period.

4. Discussion

Several scientists have already examined the role of dietary enzymes in fish development, feed consumption, nutrient utilization, and plasma production [11, 12, 4, 13, 14]. There was no information on the use of the enzyme Thepax Aqua in the diet for juveniles of *C. carpio. Still, a* few studies were done on the effect of other enzymes, such as using photolytic enzymes on this and other species. However, Tagare [15]. reported that proteolytic enzymes have an important part in protein digestion. Dabrowski and Glogowski [16]. found an increase of proteolytic activity in common carp *C. carpio* and their growth supplementing with bovine Trypsin to the fish diet. Carter *et al.* [17]. also stated good growth in *Salmo salar* was recorded by adding proteolytic enzymes to its food.

The current study indicated that the diet supplemented with 1 % of Thepax Aqua gave good growth of juvenile *C. carpio*. This could be caused by the increase in the digestion ability of juveniles by supplementing Thepax Aqua to the fish diet. Furthermore, this study also found that growth performance (WG, ADG, FCR, FCE) improved with using a diet supplemented with 1 % of Thepax Aqua. This development could result from enhanced feed eating and nutrient digestibility. It could be concluded from the study results that the level of enzyme used in the fish diet enhanced the growth of the fish. Robinson [4] previously demonstrated that supplementing fish Fingerlings with enzymes could greatly improve growth performance and feed utilization.

The results of this study corroborate Webster and Reed's [13] findings that a dietary enzyme ratio of 1.5 to 2.5 was optimal for channel catfish growth. A similar conclusion was reached in this investigation, which found that there was no statistically significant effect (P>0.05) for enzyme levels on survival of juveniles carp offered different dietary enzymes as given in Table (3). According to Dabrowski and Glogowski [16], exogenous proteolytic enzymes given to fish feed had no effect on the fish's survival. On the other hand, Tagare [15], reported the same results with common carp, which given a diet having papain and crude papain in the shape of papaya leaf.

Acknowledgement
This research was funded by Department of Vertebrates, Marine Science Center, University of Basrah.

References

1 - Nasir, N. A. (2013). Effect of replacement of fish meal by soybean on growth, survival, feed utilization and production cost of fingerlings common carp (*Cyprinus carpio* L.) reared in the float cages. International Journal of Recent Scientific Research, 4 (4), 308 -312.

2 - Nasir, N. A. & Hamed, Q. (2016). Growth development of young common carp *Cyprinus carpio* through dietary sodium chloride supplementation. Mesopotamia Environmental Journal, 2 (2), 12-18.

3 - Dabrowska, H.,; Grudniewski, C. and Dabrowski, K. (1979). Artificial diets for common carp: effect of the of the addition of enzyme extracts. The Progressive Fish-Culturist, 41, 196-200.

4 - Robinson, E. H. (1994). Effect of high protein "Fishing" feeds on performance and fat content of channel catfish. Journal World Aquaculture, 25(3), 456-470.

5 - Hoehne-Reitan, K., & Kjørsvik, E. (2004). Functional development of the liver and exocrine pancreas in teleost fish. Proceeding of American Fisheries Society Symposium, 40, 9-36.

6 - Klein, S.; Cohn, S.M. & Alpers, D. H. (1998). The alimentary tract in nutrition. In: Shils, M. E., Olson, A. J., Shike, M., Ross, A.C. (Eds.), Modern Nutrition in Health and Disease, 605-633.

7 - Lovell , R. T. (1989. Nutrition and feeding of fish. Van Nostrand rein hold, Newyourk, 260 pp.

8 - AOAC.(2000). Association of Official Analytical Chemists. 17th Edition. Hortuntzed (Ed). Washington, USA. 21-447.

9 - Nasir, N. A.; Yesser, A. T. and Al-Hamadany, Q. H. (2019). Effect of pH on the growth and survival of juvenile common carp (*Cyprinus carpio* L.) Iraqi Jounal of Science, 60 (2), 234-238.

10 - Hekmatpour, F.; Kachanin, P.; Marammazi, J. G. and Mousari, S. M. (2019). Changes in serum biochemical parameters and digestive enzyme activity of juvenile sobaity sea bream (*Sparidentex hasta*) in response to partial replacement of dietary fish meal with poultry by-product mea. Fish Physiology and Biochemistry, 45, 1-4.

11 - Bardócz, S.; Grant, G.; Brown, D. S.; Ralph, A. & Pusztai, A. (1993). Polyamines in food implications for growth and health. Journal Biochem., 4, 66-71.

12 - Govoni, J. J. ; Boehlert , G.W. & Watanabe, Y. (1986). The physiology of the digestion in fish. Environ. Biol. Fish., 16, 59-77.

13 - Webster, C. D. & Reed, E. B. (1994). Effect of dietary protein level on growth and body composition of catfish. Journal. Appl. Aquacult., 4 (2), 73-80.

14 - Balmer, J. E., & Blomhoff, R. (2002). Gene expression regulation by retinoic acid. Journal Lipids, 43, 1773-1808.

15 - Tagare, M. N. (1992). Role of papain in growth, survival and improving feed efficiency of *Cyprinus carpio*. M. F. Sc. Thesis submitted to Central Institute of Fisheries Education, Mumbai, 1-50.

16 - Dabrowski, K.. & Glogowski, J. (1977). A study of the application proteolytic enzymes to fish food. Aquaculture, 12,349-360.

17 - Carter, C. G.; Houlihan, D. F.; Buchnan, B. & Mitchell, A. I. (1994). Growth and feed utilization efficiencies of seawater Atlantic salmon, *Salmo salar L.*, fed a diet containing supplementary enzymes. Aquaculture Fisheries Management, 25,37-46.

Desalination methods and their role in solving and managing the problem of salinization in Iraqi water resources

Dr. Havan H. Salman; Zahraa S. Aswad

Ministry of water resources, Baghdad, Iraq

Corresponding author: havan.hassan68@gmail.com

Abstract. a rise in the demand for fresh water is due to population growth in desert areas where high-quality, low-salinity water is scarce has prompted a rapid development of desalination technology, which has demonstrated its ability to address as well as resolving the water shortages in many dry locations throughout the world. Despite the benefits that could be realized from desalination under the global water crisis, the majority of people, including those who work in the water sector, regard it as a novel technology due to its lack of application. Currently, Desalinated water is used by only 1% of the global total to meet their daily needs. Nonetheless, the United Nations predicts that by 2025, Water scarcity will afflict 14% of the world's population. For the reasons indicated in this study, Iraqi water sources have undergone a quantitative and qualitative deterioration, resulting in a rise in salt concentrations.

1. Introduction

Natural freshwater scarcity, along with population increase and industrial development in many nations, has resulted in worldwide stress and freshwater scarcity. Water distribution across the planet is depicted in NASA and USGS charts as Seawater makes up 97.5 percent of the world's water, with the remainder 97 percent being snow on perpetual glaciers or permafrost. As a result, Only 0.5 percent of the world's water supply is easily accessible for "direct" human usage as groundwater with low salinity or in lakes, and rivers. as shown in figure 1 [1].

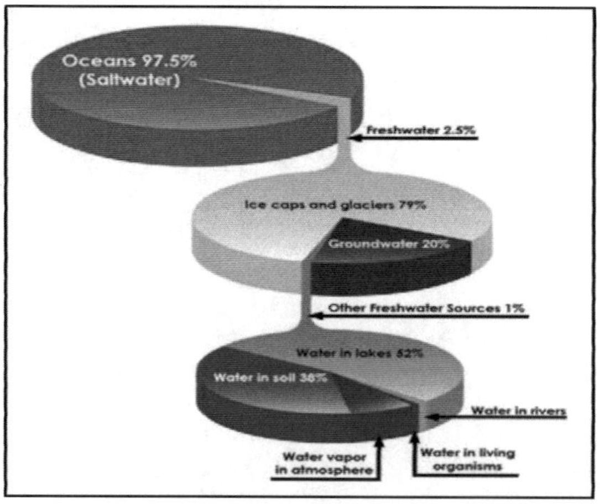

Figure 1. Global availability of freshwater

Figure 2(A) illustrates the baseline water stress for the world in 2040, according to the World Resources Institute. While figure 2 (B) shows the Water demand on a global scale (Baseline Scenario 2000 and 2050) based on; (Irrigation, Domestic, Livestock, Manufacturing, and Electricity) needs

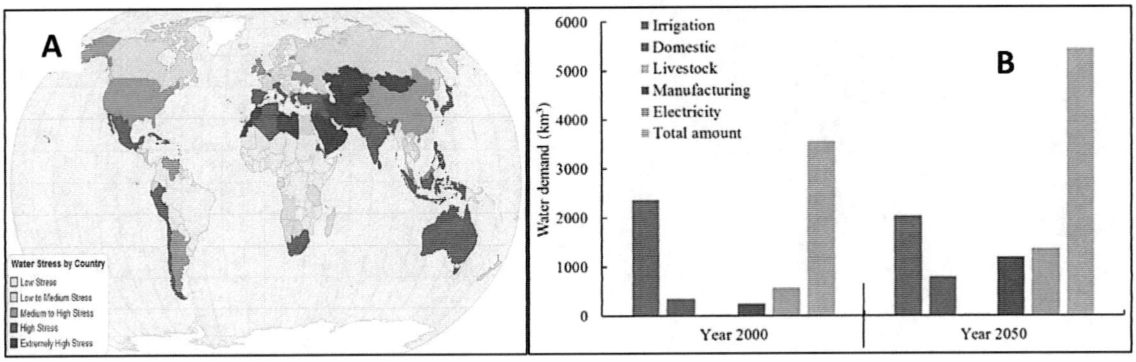

Figure 2. (A) World water stress, (B) World water demand [2]

Iraq's main source of surface water are the Tigris and Euphrates rivers, as seen in figure 3. The amounts of water in Iraq's rivers, on the other hand, in recent years, their capacity has plummeted to less than a third of what it used to be. Due to decreasing precipitation, progressive desertification, and water usage and damming in upstream countries, water levels may plummet even more in the coming years. figure 4 shows how Iraqi water is classified based on its composition and application. [2].

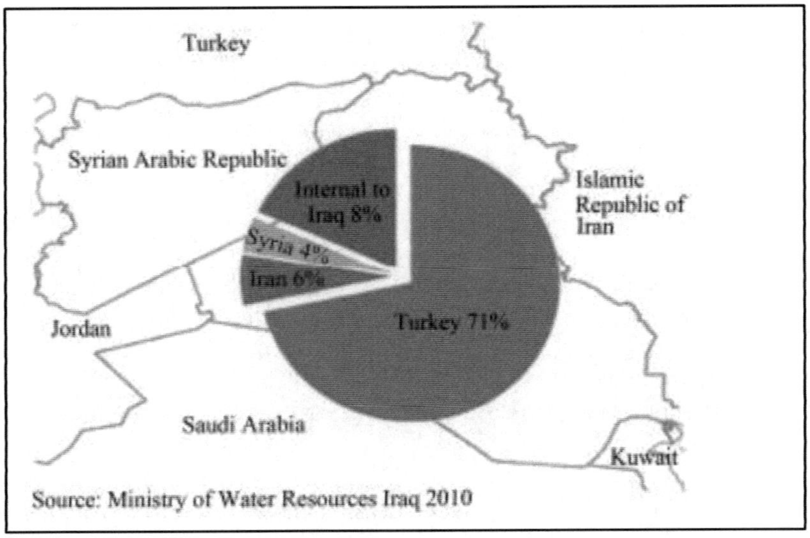

Figure 3. Iraqi water sources [MWR, 2010]

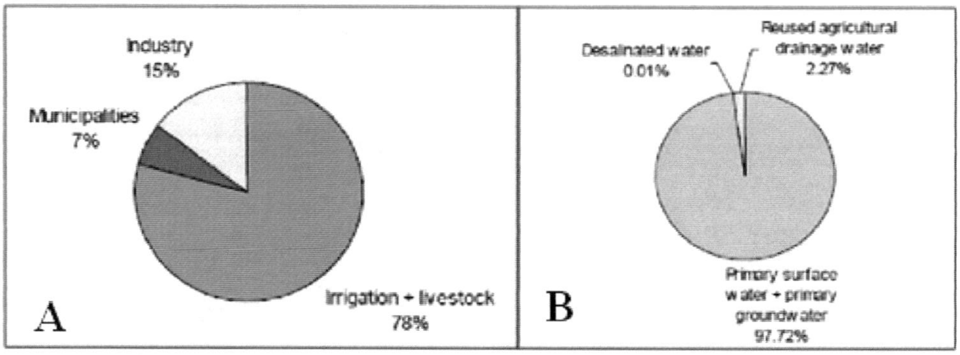

Figure 4. Iraqi water classification (A) Water withdrawal by sector, (B) Water withdrawal by source [2]

Iraq has recently had extremely lengthy seasons of drought and heat, with frequent dust storms and rising river levels at historic lows. Input of water and quality have both fallen considerably, and The salinity of the Shatt-el-Arab River has risen to alarming proportions, according to data collected through 2014 [3]. Several studies [4, 5, 6, 7] have examined the reasons and magnitude of key Upstream developmental projects on the Tigris and Euphrates that have led to the current predicament. TDS levels in Tigris water range from 280 to 275 mg/l at the Turkish-Iraqi border, and exceed 1800 mg/l in Basra [8]. The situation could be even worse in tributaries when TDS levels in the Diyala River reach 3705 mg/l. [9].

The Decay of water quality and rising soil salinity have resulted in major desertification concerns all across Iraq. Desertification is thought to have affected at least 45 percent of Iraq's land area [10]. The influence of salinity on farming systems in Central and Southern Iraq is a severe constraint to agricultural productivity. At the farm size, a strategic focus on investment in salinity management is required. [11]

Improving salinity water management in Iraq necessitates looking beyond common water technologies and considering alternative resources available to improve water security and protect these resources from upstream countries' threats of cutting or reducing them, through large investments in non-traditional sources, such as available saline water, especially in wetlands and MOD. [10]

Related to the above situation with the abundance of salty water resources, the desalination technology was found to be a highly applicable technology to address present and future freshwater demands in many parts of the world.

In1931, Ross emphasized that Aristotle, the ancient Greek philosopher, was the first to establish the concept of desalination during his study Meteorology, he noticed that "salt water, when it converts into vapor, becomes pleasant, and the vapor does not form salt water again when it condenses," and that a fine wax vessel could store drinkable water after being submerged in sea water for a long enough period of time. [12].

Desalination is the process of removing soluble salts from seawater and, in some cases, brackish (slightly salty) inland sea waters, highly mineralized groundwater, and municipal wastewaters [13].

In a larger sense, desalination is the process of eliminating salts and minerals from a target item to make otherwise useless water appropriate for human consumption, agriculture, industrial applications, and a number of other uses. [14].

2- Application of Desalination technology in Iraq

Despite the previously mentioned water issue, we discover that desalination applications in Iraq are not up to the task. According to the Iraqi Environment Survey for 2010, the number of Iraqi governorates producing desalinated water has increased to ten governorates, with the majority of the water sources coming from projects and water aggregations accounting for 89.7% and ground water accounting for 10.3%. [15].

In Table (1), it was shown that count of desalination plants in Iraq (RO) are 160 plants, most of them are located in Basrah and Dhi Qar governorates which are (49 and 48) plants, respectively, with a design capacity of 78000 m3/day whereas the amount of desalinated water reached to 9000 m3/day on which average amount of desalinated water percentage to the design capacity was 17.9%. This drop in percentage is due to the inability of 38.8% of total plants to function. [15].

Table 1. distributed of Iraq desalination plants (RO) @2010 [15].

Governorate	Number of Desalination Plants	Water Source
Kirkuk	11	Ground water
Dhyala	1	Ground water
Anbar	2	Ground water
Karbala	2	Projects and aggregations water (surface water)
Wasit	19	Ground water
Najaf	2	Ground water
Muthana	13	Projects and assemblage water (surface water)
Dhi Qar	48	Projects and assemblage water (surface water)
Maysan	12	Projects and assemblage water (surface water)
Basrah	49	Projects and assemblage water (surface water)

1- Desalination processes development

Desalination, like many other technologies, did not evolve in a linear fashion. There were times of fast expansion, yet there were also times when the company was dormant. In England, the first patent for a desalination technique was issued in 1869. The British government established the first water distillation facility in Aden the same year to feed ships calling at the Red Sea port. Around 18,000

desalination plants were in operation around the world by 2019, producing more than 95 million cubic meters in total (over 3.4 billion cubic feet) of potable water on daily basis. [16].

Desalination is particularly important in dry regions like Australia, where water has traditionally been collected behind dams. Kuwait generates more water than any other country in terms of desalination, accounting for the entire amount of water used [17]. The two basic types of desalination methods utilized worldwide can be essentially classified as thermal or membrane. figure 5 [16] demonstrates an increase in the capacity of the installed membrane and thermal desalination. To function and create fresh water, both devices require energy. There are sub-categories (processes) that use diverse methodologies within those two basic types. [18].

Figure 6 depicts a schematic representation of the family of desalination methods that are used for large-scale water production. Desalination processes can also be divided into categories based on the type of main energy source utilized to power them. Figure 7 depicts this categorisation [19].

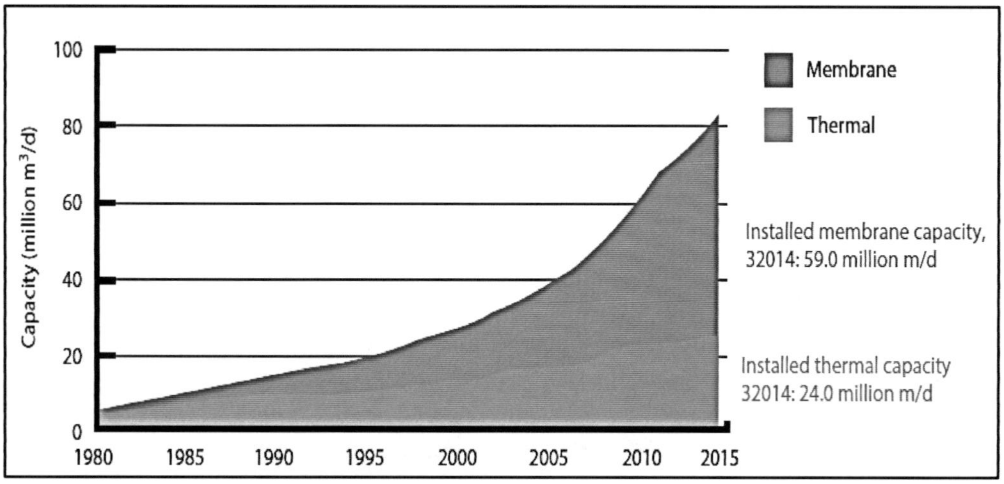

Figure 5. The expansion of installed membranes and thermal desalination [Capacity Global Water Intelligence]

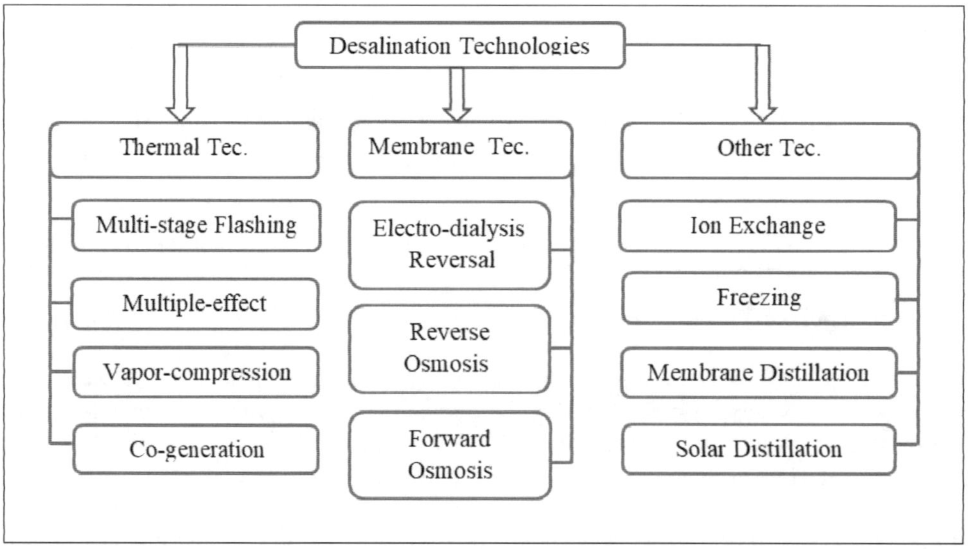

Figure 6. The most extensively utilized techniques [19]

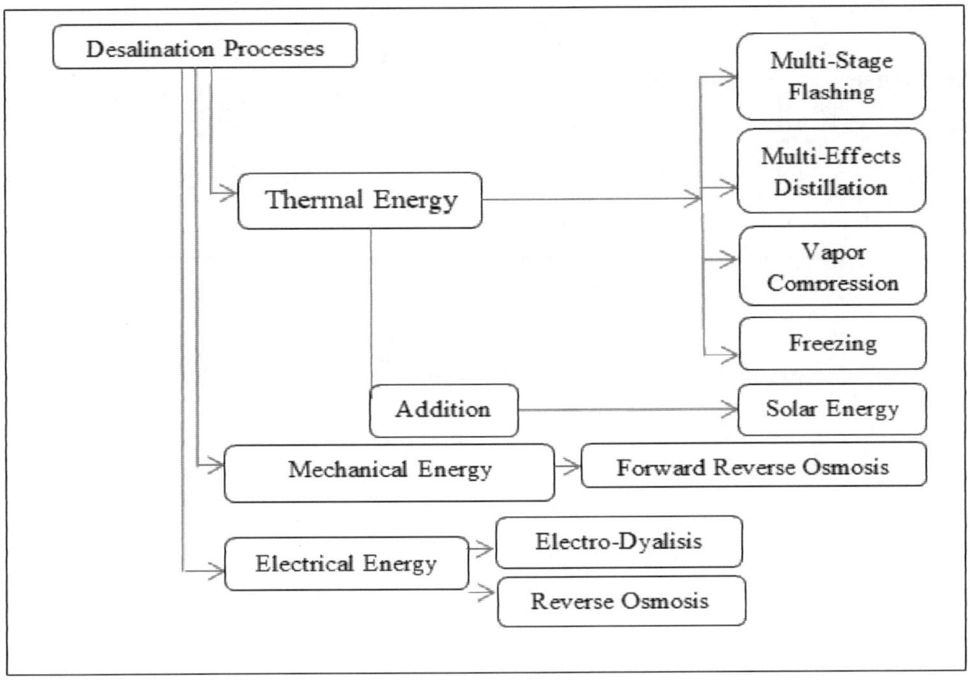

Figure 7. Desalination process energy classification [19]

In early 2000, global thermal and membrane capacity was around 7 billion gallons per day (bgd-), with around 50 percent in thermal processes and 50 percent in membrane technologies. This stands for total installed capacity from the early 1950s, nonetheless not all of it is likely to be operational. From 1972 to 1999, global desalination capacity rose at a rate of almost 12% per year. Over 8,600 desalination plants have been constructed worldwide, with the United States accounting for roughly 20% of them, the highest percentage of any country. [18].

2- Thermal Desalination (Distillation)

Distillation is the traditional desalination method [19]. Desalination by thermal means, often known as distillation, is one of the oldest methods of converting brackish water and seawater into drinkable water. Boiling, evaporation, and condensation principles are employed. The water is heated until the point of evaporation is reached. While the vapor is condensed to make fresh water, the salt is left behind [20]. Steam generators, waste heat boilers, and back-pressure steam extracted from turbines in power plants now provide the essential thermal energy. [21].

a. Thermal-based technologies,

vapor compression distillation, multi - effect distillation, and multistage flash evaporation are three techniques that have been employed in seawater desalination since the beginning and are widely used in the Middle East and the Caribbean. These technologies are generally used in areas with high salinity (> 35 gL^{-1}) and low-cost energy (e.g., Middle East, co-located power plants) [22]. Four primary commercial processes for thermal desalination have been examined here:

- **Multistage flash (MSF) distillation,** the compelled circulation technique is the top durable of all desalination systems and can produce enormous amounts of water per unit. MSF is one of the most extensively used desalination technology throughout the world; where, they use a series of chambers or stages, each with progressively lower temperature and pressure, to quickly evaporate (or flash) water from the bulk liquid.

Figure 8 shows how the vapor is condensed by inflowing feed water tubes, permitting energy to be recovered from the condensation heat. The number of stages in the MSF procedure is directly related to how well the system uses and reuses the heat it receives. [22].

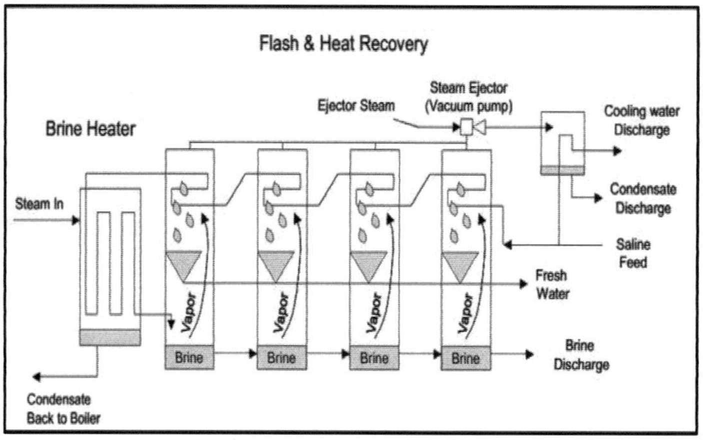

Figure 8. Multi- flash distillation [23].

- **Multi-effects distillation (MED),** figure 9 shows how (MED) operates through a sequence of phases known as "effects". figure 10 shows how (MED) works. Water from the faucet is sprayed into pipelines, which are subsequently heated in order to produce steam. The steam is then utilized to warm the next group of sea water that arrives. To increase efficiency, the steam needed to heat the sea water might be sourced from nearby power plants[22]. Despite the fact that this among heat-powered systems, it is the most thermodynamically efficient. [24], it does have some restrictions, such as a maximum temperature and maximal quantity of effects. [22].

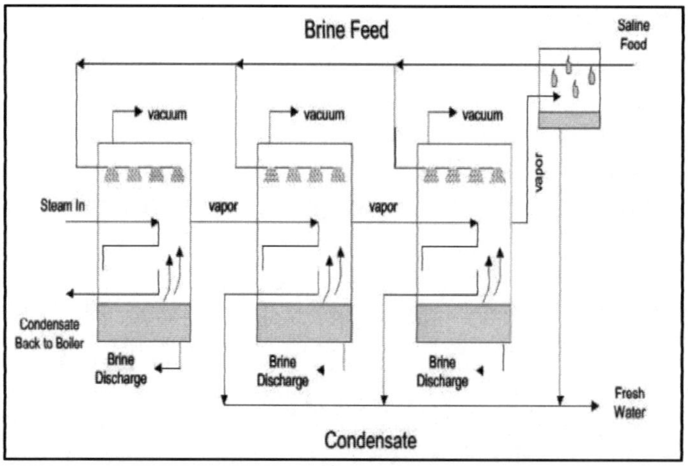

Figure 9. Multi-effects distillation [23]

- **Distillation using vapor compression (VC)**, (VC) includes evaporating the supply water, compressing the resultant vapor, and finally utilizing pressured vapor as a source of heat to evaporate more supply water. The vapor is compressed using a mechanical compressor (mechanical vapor compression, MVC) or a steam ejector (thermal vapor compression, TVC). MVC systems with only one stage can be as small as 3,000 (m^3day^{-1}), but TVC systems with many stages can be as large as 20,000 (m^3day^{-1}). This discrepancy arises from the fact that MVC systems have the same specific power consumption (power/unit water generated) regardless of the number of steps, whereas the thermal efficiency of TVC systems is improved by adding extra stages [26]. MVC is given for the most part in Figure 10 mechanical vapor compression; VC procedures are practicable for small to medium scale installations. [27].

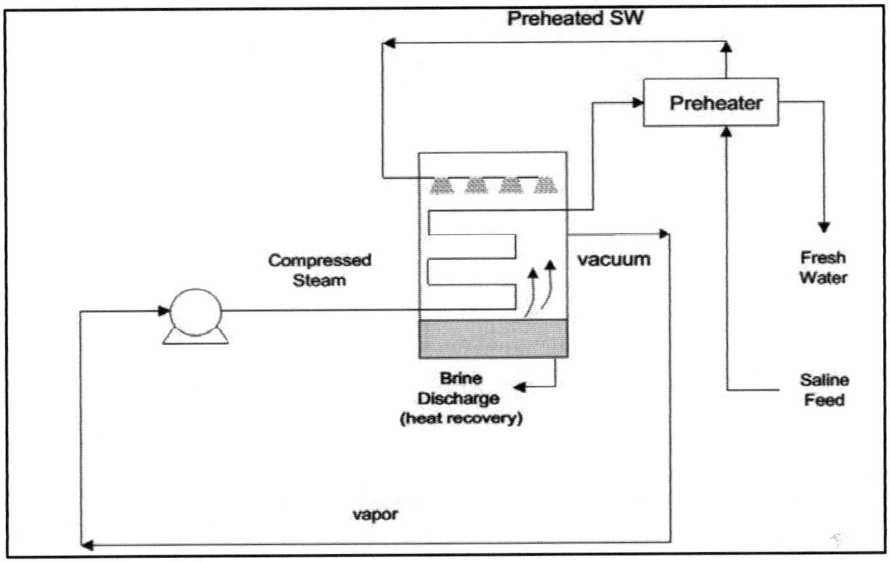

Figure 10. A single mechanical vapor compression distillation process is depicted schematically. [25]

- **Power and water desalination cogeneration system;** Energy can be employed in dual-use or cogeneration systems, where the energy sources can fulfil several functions, such as power generation and water desalination. [29]. Cogeneration plants linked to Units that deal with seawater in a multi-stage flash desalination process produce the majority of consumable desalinated water and energy in the Arabian Gulf countries and North Africa. The primary processes of a cogeneration system are: high-pressure steam is used to power turbines, which then power electric generators. In most circumstances, boilers generate high-pressure steam at roughly 540°C (1,000° F). The temperature and energy level of this steam decreases as it expands in the turbine. Distillation plants require steam with a temperature of 120°C (248°F) or less, this can be obtained by withdrawing lower temperature steam from the turbine's low pressure end after the majority of its energy has been consumed to create electricity. This steam is then passed via the brine heater at the distillation plant, raising the temperature of the entering seawater. Before being used in the turbine, the steam condensate is returned to the boiler and reheated. [29].

b. Membrane Distillation

A membrane that separates two phases permits one or more components to be transported more quickly than other components. A pressure gradient, as shown in figure 11, a gradient in temperature, concentration, or electrical potential can all be used to propel transport. Reverse osmosis (RO) and electro dialysis are the two most common membrane processes used for desalination (ED). In rare circumstances, forward osmosis (FO) is also utilized. [23].

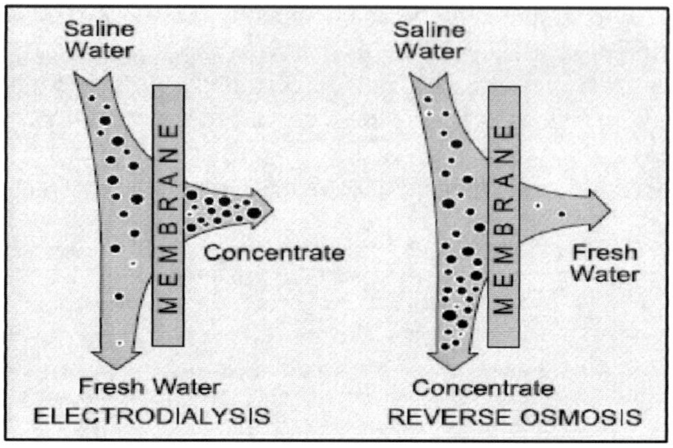

Figure 11. Driven force of (ED) and (RO) [23]

- **Electro dialysis (ED) is a technique that is powered by electricity.** An ED unit is typically composed of a series of ion-selective anion and cation exchange membranes (AEMs, CEMs) interleaved with alternating concentration and dilute solution compartments and an electrode compartment on both sides. After the electrodes have been electrically charged, an electrical-current flows through the ED stack (by connecting them to a power source). As a result, electrically charged solutes must travel in accordance with the electric field. Anions (positively charged) travel to the anode, while cations migrate to the cathode (negatively charged). The existence of IEMs, which act as optimally selective ion barriers, ensures that anions travel readily through the nearest AEM while being inhibited by the nearby CEM. Cations, on the other hand, flow via CEMs and are inhibited by AEMs, resulting in a depletion of salt content in the dilute compartments and an enrichment of the concentrate one. [30], illustrated in figure 12.

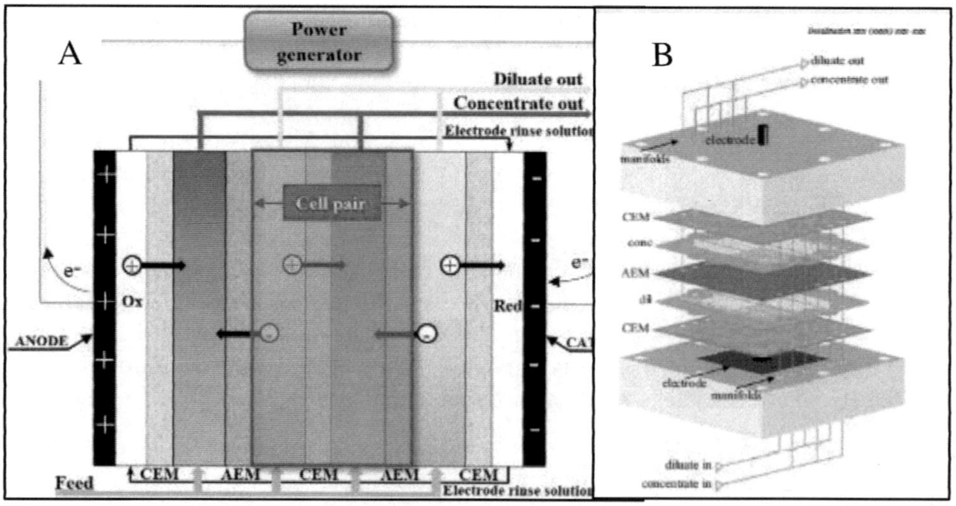

Figure 12. (A) Schematics of an ED system, (B) ED stack in the sheet flow arrangement of a laboratory scale unit with net spacers [30].

- **Reverse osmosis (RO)** is a membrane separation process that involves passing a pressurized saline solution across a water-permeable membrane to separate water from dissolved salts figure 13. The pressure differential formed between the pressurized feed water and the product water, which is close to atmospheric pressure, supports the liquid moving through the membrane to do so. The remaining feed water is converted to brine as it passes through the pressurized side of the reactor. There is no heating or phase change. The initial pressurization of the feed water necessitates a substantial amount of energy. The operating pressure for brackish water systems ranges from 15- 25 bar; whereas for seawater systems a range from 54 to 80 bars (the osmotic pressure of seawater is about 25 bar) [27].

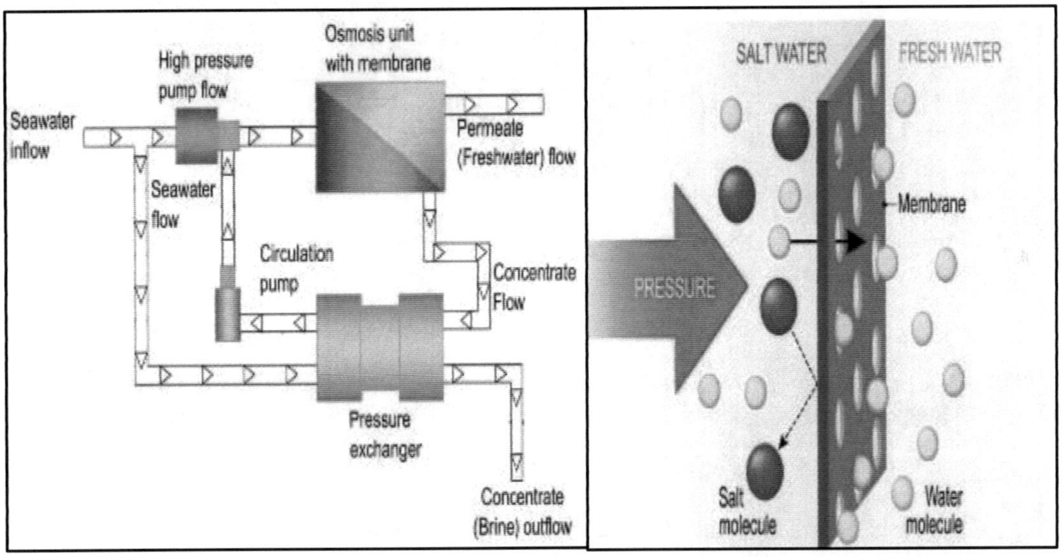

Figure 13. Schematics of RO system & sheet flow arrangement [31].

- **Forward osmosis (FO),** is a new membrane technology that could be used to clean water in a variety of ways. The feed and draw solutions (DSs) chosen, as well as the water quality objectives, determine the potential FO uses. The process can be driven with almost no external hydraulic pressure. Water is extracted from a lower osmotic pressurized feed solution (FS) into a higher osmotic pressure feed solution (DS) in FO; the process is facilitated by the osmotic pressure difference between the two aqueous solutions on opposite sides of the semipermeable membrane, resulting in the FS being concentrated and the DS being dissolved. The osmotic pressure gradient formed by a highly concentrated DS to let the water to flow through a semi-permeable membrane in the FO process comes from FS, which has a lower concentration figure 14. As a result, a less concentrated draw solution is created, which can then be processed further to remove freshwater. With the application of a suitable draw solution, high osmotic pressure driving forces can be created, resulting in high recoveries and, in theory, salt precipitation. [33].

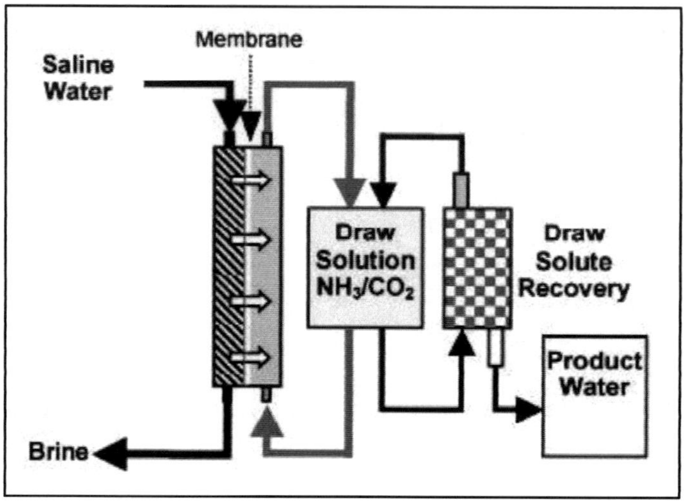

Figure 14. Schematics of Forward Osmosis Process [32].

c. Other Desalination Processes

Desalting saline waters has been accomplished using a variety of methods. These procedures haven't had the same commercial success as distillation, ED, and RO, although they could be useful in certain situations or with future research.

- **Freezing:** In the 1950s and 1960s, a lot of effort was put on developing freezing desalination. Dissolved salts are naturally eliminated during the production of ice crystals during the freezing process. Under controlled regulations, seawater can be desalinated by cooling saline water and forming ice crystals.. Before the full volume of water is frozen, it is normally cleaned and rinsed to remove any salts that may remain in the leftover water or adhere to the ice crystals. After that, the ice is melted to provide fresh water [27]. As shown in figure 15,the device that takes heat from the brine to form ice crystals that can be easily transferred, removed, separated, cleansed, and melted differs between the various freezing methods. Indirect freezing and direct freezing procedures are the two types of freezing processes. Secondary refrigerant freezing and vacuum freezing are examples of the latter. The essential points of these processes are outlined in the following paragraphs. [34].

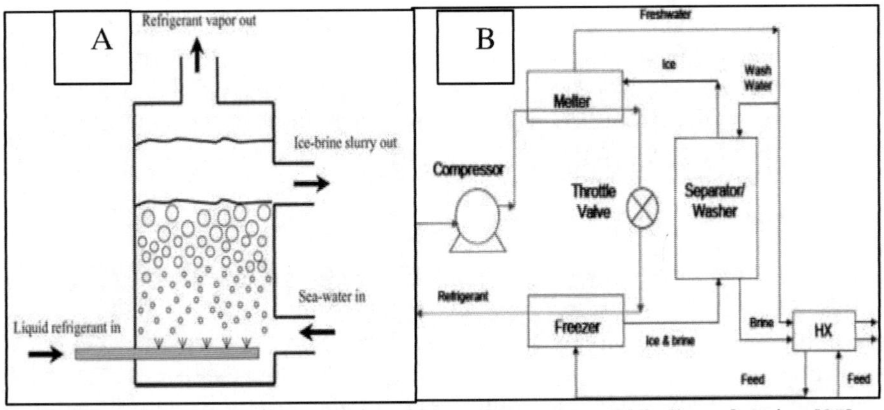

Figure 15. Freezing Desalination; (A) direct freezing processes,(B) indirect freezing [27]

- **Membrane Distillation**, was introduced to the market on a modest basis in the 1980s, but it has yet to become economically successful. The technique employs both distillation and membranes, as the name implies. The procedure involves warming saline water to increase vapor generation, then exposing the vapor to a membrane that can pass water vapor but not liquid water. After passing through the membrane, the vapor condenses on a cooler surface, yielding fresh water. Fresh water cannot travel back through the membrane in its liquid form, thus it is retained and collected as the plant's output [27]. To induce a vapor pressure difference across the membranes for MD, four different approaches have been used [35]. In figure 16, it is shown the four methods are:

1- 1- Make a direct contact MD, MD's most basic configuration is this one. In this configuration, vapor from a feed stream passes across the membrane and condenses into a solution running on the permeate side. Heat transmission from the heated feed to the product (temperature polarization) might be a problem if temperature is employed to create the vapor pressure differential.

2- Air gap MD, in this scenario, an air gap separates the hydrophobic membrane from the cold condensing surface This is one of the most adaptable techniques .

3- Gas sweeping The MA flowing gas is utilized to sweep the vapor off of the membrane permeate side, ensuring that the transport gradient is maintained. This is especially beneficial for eliminating unstable or dissolved gases from liquid streams.

4- Vacuum MD, A vacuum is maintained on the permeate side of the membrane to facilitate vapor passage across the membrane. This is also beneficial for extracting volatile components or degassing liquids.

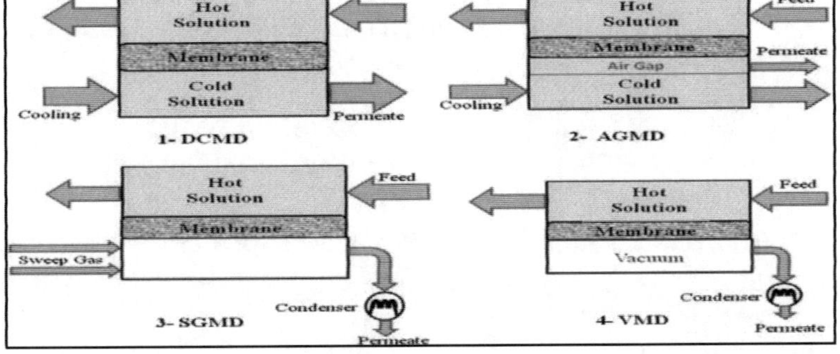

Figure 16. Methods employed in a membrane distillation scheme [36]

- **Solar distillation;** As shown in figure 17, Solar radiation passes through the clear glass or plastic cover, heating and evaporating the salty (or brackish) water in the basin. On the inside of the glazing, water vapor condenses, and the liquid streams into a trough, where it is collected by gravity. To improve solar absorption, basins are painted black, and long wavelength energy cannot flow through the glazing from the solar still. To put it another way, the greenhouse effect makes solar still appear to be a heat trap. A solar still needs to be flushed to avoid salt deposition, and the frequency of flush is determined by the quality of the influent [37]. Paton clear that there are two types of solar desalination [38]:-

1- Use photovoltaic cells to fuel the desalination process; therefore, solar energy is converted to electrical energy.
2- The Use of solar energy in the form of heat; is referred to as solar thermal powered desalination.
3- The natural water cycle that the Earth goes through is solar distillation.

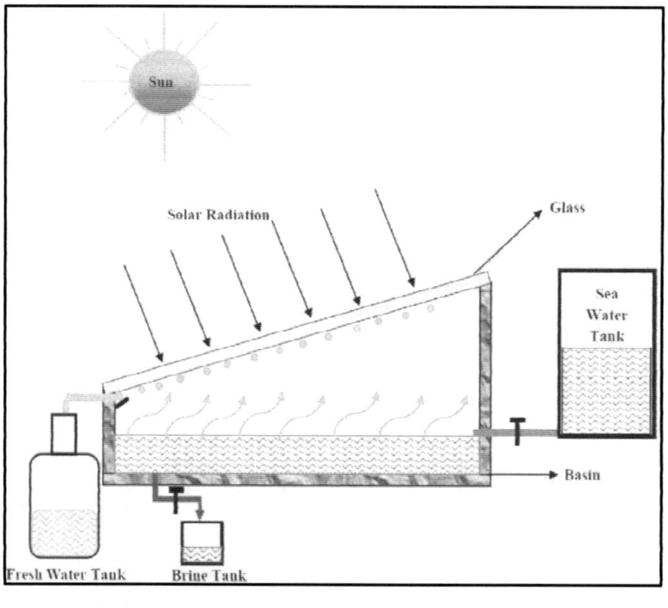

Figure 17. solar distillation natural water cycle [39]

IOP Conf. Series: Earth and Environmental Science 1002 (2022) 012004 doi:10.1088/1755-1315/1002/1/012004

Table 2. The Benefits and Drawbacks of Desalination Methods

	Advantages	Disadvantages
	Distillation (Thermal Desalination)	
Multi-stage flashing (MSF) [41, 42, 43]	- High-scale designs - Assured, robust technology with longevity. - Instead of boiling, flashing decreases the chance of scaling. - Feed water must be pretreated as little as possible. Water of superior grade - The cost and procedure of the plant are unaffected by saline levels. Heat energy can be obtained by mixing it with other sources of energy. - power generation	- Huge investment in capital is necessary Energy-intensive process - Larger footprint required - Corrosion issues if low-quality materials are employed - Slow startup rates (hours). - Maintenance often requires whole plant to shut down - Technical expertise is necessary at a high level. - Recovery ratio low
Multi-effects distillation (MED) [43]	- Feed water must be pretreated as little as possible. - Very efficient process with few operational staffing level requirements. - Normal levels of suspended and organic debris are tolerated. - Heat energy can be generated by mixing it with electricity generating. - Extremely high-quality product water	- High energy consumption - High capital and operational cost - Because the procedure is prone to corrosion, high-quality materials are required. - Prior to being used for potable water, product water must be cooled and blended.
Vapor compression (VC) distillation [8, 23]	- The plant's functioning is simple and reliable. - Used in resorts, industry, and drilling locations where fresh water is scarce. - Operating temperature and power consumption are low. - Scale formation and tube corrosion are reduced as a result of this. - The plant's initial cost is modest, and its operation is straightforward and reliable. - In comparison to RO, the plants can create high-quality water from low-quality input water.	- Large, costly steam compressors are required. - Because the evaporator components are immediately exposed to the feed water, scaling and corrosion are important concerns. (100)
Cogeneration system	- can significantly reduce the consumption of fuel	- the units are permanently connected together - The steam turbine has to be turned on. - This persistent linkage has the potential to be problematic. - when the demand for power is low or when the turbine or generator is out of commission for maintenance.
	Membrane Desalination	

Electro-dialysis reversal (EDR) [44]	- The amount of energy used is proportional to the number of salts removed rather than the amount of water treated. - A longer membrane life of 7–10 years is possible. - At low to moderate pressures, it can be used.	- Only appropriate for feed water with a TDS of up to 12,000 mg/L. - Membrane cleaning is necessary on a regular basis. - Membrane stacks are susceptible to leaks. - Bacterial pollutants are not eliminated by the system - Potable water necessitates post-treatment.
reverse osmosis (SWRO) [45]	- Investing costs are relatively cheaper. - There is no flow of cooling water. - Simple operation and quick boot-up - High-capacity manufacturing footprint - Recoveries of 30–60% - Other pollutants were removed in addition to salts. - Modular construction.	- Increased chemical and membrane replacement expenses - Weakness to feed changeable- quality water - A proper pretreatment is required. - Membranes prone to biofouling - High-pressure mechanical failures - A minimum membrane life expectancy of 5–7 years is expected.
Forward osmosis (FO) [46,47]	- Hydraulic pressure is low or absent. - Wide spectrum of pollutants are rejected with high efficiency. - Reduced membrane fouling proclivity. - membrane activities that are triggered by pressure - Fouling is minimal, and energy is used sparingly.	- influenced by the chemical characteristics of the draw solute. - physical structure - Draw solute regeneration from diluted draw solutions - The generation of clean water may need a significant amount of energy. - Low-performing FO membranes
	Other Desalination Processes	
Ion exchange (IEX) [48]	- Electrical energy is the only source of power. - Effectively removes boron	- Low-concentrated salts - Seawater is not marketed.
Freezing	- For single-stage operation, the theoretical energy need is smaller. - lower risk of corrosion. - There aren't many scaling or precipitation issues. - Can be used to remediate industrial pollutants more effectively than drinking water.	- Involves moving and processing mechanically difficult-to-move ice and water mixtures. - The procedure has not been commercially successful in producing fresh water for municipal use.
Membrane distillation [49]	- There is no pressure applied. - large capacity for rejection - Vacuum operation at low temperatures - To avoid corrosion, plastic material might be used. - The salt content in the feed has little impact on performance. - Waste heat, solar energy, and geothermal energy are all viable options.	- In bench-scale or pilot-scale research - There are no membranes or modules built specifically for MD. - Membrane fouling when the membrane is moist - feed water source must be pretreated.

Solar distillation	[8, 50]	- Cost-effective for small-scale manufacturing for households and small groups, particularly where solar energy and low-cost workforce are plentiful.	- Solar collecting necessitates a vast area. - Expensive installation. - susceptible to weather damage. - Expensive labor

3- Conclusions

This study clarified the global water crisis and its impact on the deterioration of fresh water resources in Iraq, which coincided with the upstream nations' expansion and development of storage and irrigation projects that depleting these resources.

The Mesopotamian water has suffered in both quantitative and qualitative degradation, leading to an increase in salt concentrations of the Euphrates River and the Shatt al-Arab, in addition to the local saline sources represented by the water of the marshes and the waters of the MOD which necessitated the development of solutions and management plan in order to control this deterioration and to exploit the high salt concentrations to produce and cover the need of fresh water through the application of desalination techniques, with the presence of the factors that assist the successful application of this technology:

- The marine site of Iraq, which overlooks the Arabian Gulf with a coastline of 60 km long, provides an adequate amount and covers the region's need of raw water prepared for desalination.
- The high salt concentrations of the Euphrates River and the surfaces of the Iraqi marshes make them an important source of raw water for desalination purposes
- Downstream water and other agricultural drainages is also an enhanced and alternative source of fresh water after desalination.

4- Recommendations

1. Desalination technology is a suitable solution to the water crisis in Iraq especially in the southern part, where salted water is considered as a problematic phenomenon.

2. Strengthening the regional relations of Iraq in the field of Desalination and benefiting from their experiences especially the countries of the Gulf Basin.

3. Finding an environmentally friendly and economic solution to the fate of the return of the desalination process, this can be done by;

    a. Dumping highly saline water into the sea offshore or, Mixing it with sewage water after treatment, or with cooling water from power plants to reduce the degree of salinity before dumping it into the sea.

    b. Dried and salt produced.

    c. Apply the Solvay new technology (Get rid of carbon dioxide and concentrated brine by converting it to bicarbonate instead of chlorides and thus produce soda) which it under research.

4. Adopting current designs for desalination plants for marine waters, which should be acceptable for the quality of coastal waters due to their high total dissolved salts content.

5. Activating domestic desalination plants especially solar powered to ensure that the population's needs are met.

References

[1]Pbslearningmedia,https://www.pbslearningmedia.org/resource/buac17-35-sci-ess-waterdistribute/ saltwater and freshwater distribution on earth /

[2] World Resources Institute, 27 Aug 2015, retrieved from Water Stress by Country: 2040 - World | Relief Web

[3] Karim U, Augustijn D, Waisi B, Hulscher S 2016 Evaluation of water demand and supply in the south of Iraq *Journal of Water Reuse and Desalination*, vol 6 pp 214–226

[4] Unver I, Olcay H 1997 Southeastern Anatolia Project (GAP) *International Journal of Water Resources Development* vol 13 pp 453-484

[5] Yuksel I, 2006 Southeastern Anatolia Project (GAP) for Irrigation and Hydroelectric Power in Turkey *Energy Exploration & Exploitation*. Vol 24 pp 361-370.

[6] Tech Environment (UK)LTD, Package Water Treatment Plant.

[7] Jones C, Sultan M, Yan E, Milewski A, Hussein M, Al-Dousari A, Al-Kaisy S, Becker R, H. J. 2008 Hydrologic impacts of engineering projects on the Tigris-Euphrates system and its marshlands *J. Hydrology* vol 353

[8] Ali Al-Saffar E, & Al-Joboury Z J 2017 Identifying Factors Affecting Water Resources of Iraq by Application of Knowledge Discovery In Databases. *Al-Qadisiyah Journal for Engineering Sciences*, vol 10 pp 319-337.

[9] Al-Ansari N, Ali A, and Knutsson S 2014 Present Conditions and Future Challenges of Water Resources Problems in Iraq. *Journal of Water Resource and Protection* vol **6** pp1066-1098.

[10] Al-Ansari N 2013 Management of Water Resources in Iraq: Perspectives and Prognoses *Engineering* Vol. 5 pp. 667-684.

[11] Al-Dabbas M 2013 Managing Salinity in Iraq's Agriculture, Ed W. E. Christen and K. A. Saliem

[12] Abid I, Mohammed G. 2020 Map of Locations and Values of Salt Concentration in Marshes *Ministry of Water Resources/ National Center for water Resources Management/ Department of Geographic Information Systems and Remote Sensing*

[13] Khordagui H 1999 *Ency. of Earth Sc.,* (Springer, Dordrecht). PP 124-125

[14] In: Environmental Geology. Britannica, The Editors of Encyclopaedia. "Desalination". Encyclopedia Britannica, 26 Sep. 2019, https://www.britannica.com/technology/desalination. Accessed 26 March 2021

[15] Hidalgo A, León G, Gomez M, Murcia M, Gomez E, and Gomez J L 2013 *Application of the Spiegler-Kedem-Kachalsky model to the removal of 4-chlorophenol by different nanofiltration membranes*. (Desalination vol 315)

[16] Environmental Survey in Iraq: Water, Wastewater, and Municipal Services. Detailed Report, 2011, (Ministry of Planning/ Central Statistical Organization).

[17] Kucera J. 2019 J. Desalination. 10.1002/9781119407874, 1–49.

[18] Chojnacka, K, Witek-Krowiak A, Moustakas K, Skrzypczak D, K.Mikulaa, Loizidou M. 2020 A transition from conventional irrigation to fertigation with reclaimed wastewater: Prospects and challenges *J. Renewable and Sustainable Energy Reviews*, vol 130

[19] Hisham T and Ettouney H M, 2002. Fundamentals of Salt Water Desalination (Elsevier Science) pp 690

[20] Eyvaz, M.et al 2018. J. Desalination of Water".72352(Chapter16).

[21] Winter T, Pannell D J, McCann L M J 2002 *The Economics of Desalination and It's Potential Application to Australia* (Australian Agricultural and Resource Economics Society, Canberra, Australia) 125611

[22] Raluy R, Serra L, Uche J, and Valero A. 2004 Life-cycle assessment of desalination technologies integrated with energy production systems *Desalination* vol 167 pp 445-458.

[23] National Research Council 2008 *Desalination* (The National Academies Press Washington DC) 978-0-309-11923-8.

[24] Islam M S, Sultana A, Saadat A H M, Islam M S, Shammi M, & Uddin M K 2018 Desalination Technologies for Developing Countries: A Review *Journal of Scientific Research* vol 10 pp 77–97.

[25] Paton C and Davies P 2006 *The seawater greenhouse cooling, fresh water and fresh produce from seawater* (The 2nd International Conference on Water Resources in Arid Environments, Riyadh).

[26] James E 2003 *Review of Water Resources and Desalination Technologies* Sandia National Laboratories, Report number: SAND2003-0800

[27] El-Dessouky H T, Ettouney H M, Al-Roumi Y 1999 Multi-stage flash desalination: present and future outlook *Chemical Engineering Journal* vol 73 pp 173-190

[28] Buros O K 2000 *The ABCs of desalting* International Desalination Association (IDA), Saline Water Coversion Corporation (Denver), p 30

[29] Paul F, Ramesh T, Ashleigh C 2017 Water production through CO2 capture in coal-fired power plants *Energy Science & Engineering* vol 5 pp 244-256

[30] Cath T Y, Childress A E, and Elimelech M 2006 Forward osmosis: principles, applications, and recent developments *Journal Membr. Sci.* vol 281 pp 70–87.
[31] Campione A, Gurreri L, Ciofalo M, Micale G, Tamburini A, Cipollina A 2018 Electrodialysis for water desalination: a critical assessment of recent developments on process fundamentals, models and applications *Desalination* vol 434 pp 121-160
[32] Mau S T and Sami M 2014 Introduction to Civil Engineering: A Student's Guide to Academic and Professional Success Paperback
[33] Raphael S 2008 Energy Issues in Desalination Processes *Environ. Sci. Technol* vol 42 pp 8193–8201
[34] Tzahi Y. 2005 Direct osmotic concentration as pretreatment for reverse osmosis *Journal of membrane science* vol 257 pp 85-98
[35] Lu Z and Xu L 2010 Freezing Desalination Process Encyclopedia of Desalination and Water Resources vol 2
[36] Evans L and Miller J 2002 *Sweeping Gas Membrane Desalination Using Commercial Hydrophobic Hollow Fiber Membranes* (Sandia National Laboratories) SAND2002-0138
[37] Alkhudhiri A, Darwish N, Hilal N 2013 Produced water treatment: application of air gap membrane distillation *Desalination* vol 323 pp 2–7.
[38] Lienhard, J. H. et al 2002 J.Begell House Inc. ISSN: 1049–0787
[39] Chen C and Qin H 2019 A Mathematical Modeling of the Reverse Osmosis Concentration Process of a Glucose Solution *Process* vol 7 p 251
[40] Sharshir S W, Kandeal A W, Ismail M, Abdelaziz G B., Kabeel A E, Ynag N 2019 Augmentation of a pyramid solar still performance using evacuated tubes and nanofluid: Experimental approach *Applied Thermal Engineering* vol 160 p 113997
[41] Vassilis B, Soteris K E. Delyannis 2016 *Thermal Solar Desalination* vol 1 pp 283-326
[42] Doriano B, Fabio L M, Ngai Y Y 2018 Thermodynamic analysis and energy efficiency of thermal desalination processes *Desalination* vol 428 pp 29–39.
[43] Mabrouk A N, Fath H E S 2015 Technoeconomic study of a novel integrated thermal MSF–MED desalination technology *Desalination* vol 371 pp115–125.
[44] Gude V G, Khandan N N, Deng S 2010 Renewable and sustainable approaches for desalination *Renewable and Sustainable Energy Reviews* vol 14 pp 2641–2654.
[45] Chehayeb, K M., Lienhard J H 2017 "Entropy Generation Analysis of Electrodialysis *Desalination* vol 413 pp 184–198.
[46] Eltawil M A, Zhao Z, Yuan L 2009 A review of renewable energy technologies integrated with desalination systems *Renewable and Sustainable Energy Reviews* vol 13 pp 2245–2262
[47] Babu K J, kumar T R, Yoo D J, Phang S M, and kumar G G 2018 Electrodeposited Nickel Cobalt Sulfide Flowerlike Architectures on Disposable Cellulose Filter Paper for Enzyme-Free Glucose Sensor Applications *ACS Sustainable Chem. Eng.* vol 6 p 7.

[48] Cath T Y, Childress A E & Elimelechal M 2006 Forward osmosis: Principles, applications, and recent developments *Membrane Science* vol 281 pp 70–87.
[49] Tado K. et al 2016.J.Desalination, 378, 60–66.
[50] Novalin S, Wendland M, Kulbe K D and Fischer J salination of whey by electrodialysis and Ion Exchange Resins: Analysis of both Processes with regard to Sustainability by Calculating their Cumulative Energy Demand *Membrane Science* vol 210 pp 91–102.

Study Kinetic Reaction and Removal of Indigo Carmine Dye in Aqueous Solutions by Direct Electrochemical Oxidation

Karrar H. Kazm[1] and Sarmad T. Najim[2]

[1]Technical Engineering Building and Construction Department, Islamic university, Iraq
[2]Chemical Engineering Department, Al-Nahrain University, Iraq

E-mail: karrarhayder@iunajaf. edu. iq

Abstract

In the present work, it has been proved that the direct anodic electrochemical oxidation method is effective for removing indigo carmine dye through an application on simulated textile effluent confirmed by tests of UV-visible and COD. The effect of several factors such as current density (3,6,9) mA /cm^2 electrolyte temperature (20,30,40,50)°C, and initial coloring concentration (250,500,1000) ppm ,area ratio of electrode (anode/ cathode) (1:1) and electrode spacing (0.5,1,1.5)cm) were investigated using a well design electrochemical cell. The experiment shows that the optimum condition for removing indigo carmine dye from simulated textile wastewater are current density (9 mA /cm^2), temperature (40°C), concentration (250 ppm), electrodes distance (0.5 cm) and power consumption (3.238 w.h/m^3) with removal percent (61.7%). pseudo-first-order the reaction was most to statistically appropriate describing the machine direct anodic electrochemical oxidation of indigo carmine.

Keywords: Direct electrochemical oxidation, Indigo Carmine, Decoloraziztion, Removal.

1. Introduction

For life on our earth, water is important. This is such an essential tool because without it no living organism can live. Therefore, there is a demand for clean, unpolluted water insubstantial supply[1]. Human activities are constantly affecting our water supplies. The supply of water continues to decrease due to the depletion of resources and pollution and demand rise rapidly due to population growth, industrialization, and urbanization[2][3]. Industries use approximately 10,000 different dyes and more than 0.7 Mt is produced worldwide annually[4] [3] [5]. During the dyeing and finishing processes, the textile industry absorbs large quantities of water. Approximately 1000 m^3 of wastewater supports typical medium- textile factories every day. Roughly 30 percent of the applied dyes are expected to remain unfixed and released into the effluent [6].Wastewaters with coloring are toxic to the environment because coloring compounds are stable could be the reason for cancer [7]. Because of existing aromatic bands that have high stability, biological treatment is often ineffective [8]. The chemical structure of Indigo Carmine that shown in figure (1) is a dark blue dye mainly used in the textile industry for the dyeing of polyester fibers and denim (blue jeans). Various techniques, including adsorption, photochemical and electrochemical methods, have been recommended to extract Indigo Carmine and other dyes from wastewater [6]. The electro-chemical process has been generally recognized in recent years as a strong alternative to remediation of wastewater. This provides innovative solutions to mitigate emissions issues caused by industrial waste.

However, there are only a few papers on dye electrochemical degradation that have been published. Shilpa.R and Charan Kumar H.C conducted research on removing Indigo Carmine using Pd/graphite and published a paper on removing Indigo Carmine using Rd/graphite [7] [9]. While others researchers like Salah Ammar and Ridha Abdelhedi studied the removal of boron-doped diamonds. Finally, we study the impact of many factors like current density, the temperature of electrolyte, the concentration of indigo carmine, and kinetics reaction by using graphite as an electrodes [10].

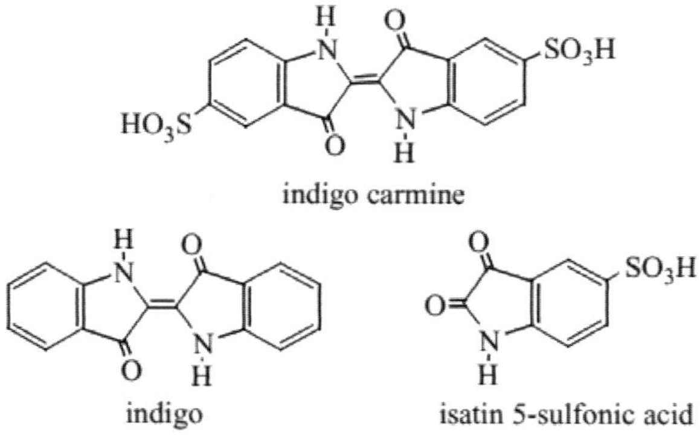

Figure (1) chemical structure of Indigo Carmine[10].

1.1 Electro-chemical technology

Several books and articles cover the application of Electrochemical to reduce environmental pollution. Electrochemical is a new technology that treatment effluent and waste management, also the environmental protection integral in the process[11][12]. The most common electrochemical procedure to remove organic pollutants is electrical oxidation or electro (EO). This technique has recently been used for the removal or degradation of aqueous dyes[13] [10] . Electrochemical is a chemical reaction that occurs to pollutants on electrodes were when the reaction happens on anode called oxidation and when occurs cathode called reduction [14] [15].

Generally, there are two techniques to electrochemical reaction:

1.1.1Direct chemical reaction

Electrochemically, the generation of $(OH^\bullet)$ in the electrolysis process occurs on the surface of the electrode (M) where the water must be activated by the electrolytic discharge of water [16] [17].

$$H_2O + M \rightarrow M(OH^\bullet) + H^+ + e^-$$
(1)

$$M(OH^\bullet) \rightarrow M + ½\, O_2 + H^+ + e^-$$
(2)

Organic compound(R) oxidation electrochemically by absorbed $(OH^\bullet)$ from the surface of the anode [18].

$$R(aq) + M\,(OH^\bullet)_{\frac{n}{2}} \rightarrow M + + \text{Oxidation product} + n/2\, H^+ + e^-$$
(3)

n: is electron number of the organic compound in a reaction

The mechanism of direct oxidation usually happen through[19]:

- Movement the contaminants from liquid to the surface of the anode.
- Oxidation the contaminates on the electrode.

Generally, The electron's movement depending on its activity and current density.

There are typically two distinct paths of anodic oxidation of organic compounds [20]:

- Electrochemical conversion: where the happens oxidation partially to pollutant as shown in below

  $$R \rightarrow RO + e^-$$
  (4)

- Electrochemical Incineration (combustion): where the reaction of pollutant oxidation transforms to H_2O and CO_2

  $$R + OH^\bullet \rightarrow CO_2 + H_2O + \text{Salts} + e^-$$
  (5)

1.1.2 Indirect chemical reaction

Through indirect electrochemical reaction will be dissociated organic pollutants by using an active agent. The most intermediate that used is chorine that obtain from the oxidation of chloride on the anode. Formation of chlorine consider responsible for the oxidation of the pollutant in bulk solution. Generally, an anode reaction for chloride will remove the impurities from the solution due to produces chlorine/hypochlorite that will be the exists or add to the aqueous solution [21] [22]. Where figure (2) shows the reaction of direct and indirect oxidation in electrochemical.

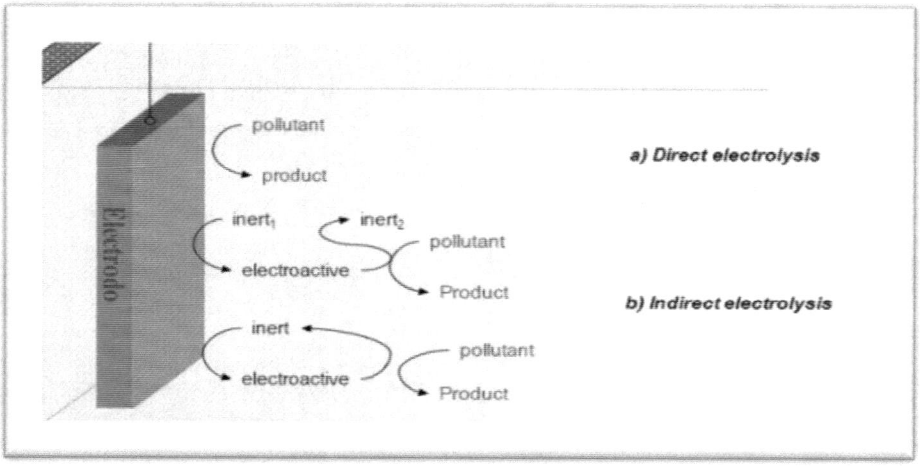

Figure (2) direct and indirect electrochemical oxidation [23].

2. Experimental

The equipment used in this experiment are the power supply, electrode (graphite), water bath, distilled water, organic pollutant (indigo carmine), Ammeter, cell, Micropipette. The Stoke solution (350mL) papered from dissolve dye (Indigo Carmine) in distilled water. Used micropipette to take 5 mL a sample from solution after each 15 min for 120 min to measure by UV- spectrophotometer and chemical oxygen demand (COD). Recording the various in an ammeter to account for the power consumption through experimental. The volume of pollutant 350 mL in the cell and the submerged part of the graphite is 9.615 cm width with length 5.2 cm. All the chemical products used were pure and approved. Different concentrations of indigo carmine with different current and temperatures were tested, PH was constant during the experiment.

The design of the cell consists of the cover that content the hole in corner and spacing between anode to the cathode is 5 mm and the high of the cell wall 10 cm from each side. As shown in figure (3, 4).

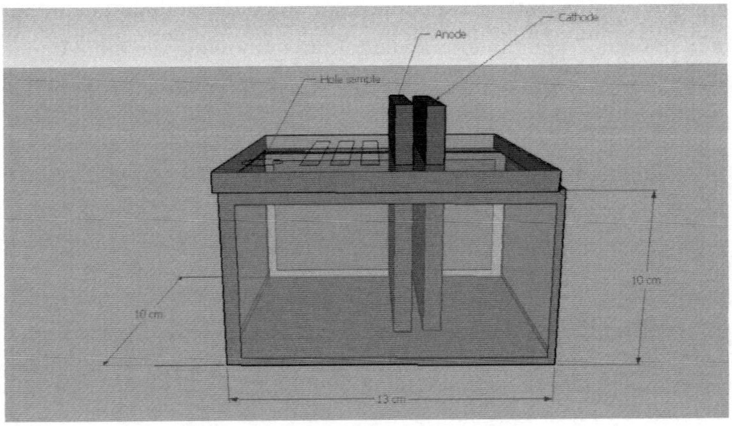

Figure (3) Design of cell reaction with the electrode.

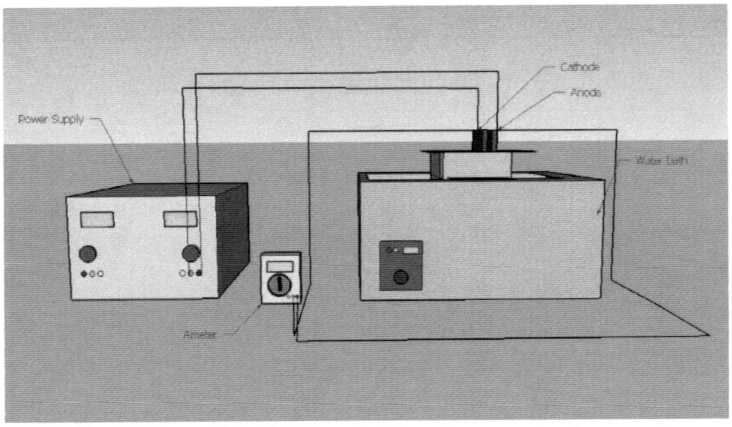

Figure (4) Design of unit operation.

3. Tests

We used the UV-spectra photometer (JENWAY 6800UV/VIS. Spectrophotometer, Britain) as shown in Figure (5) for the solution decoloration test and the COD for the percent removal test.

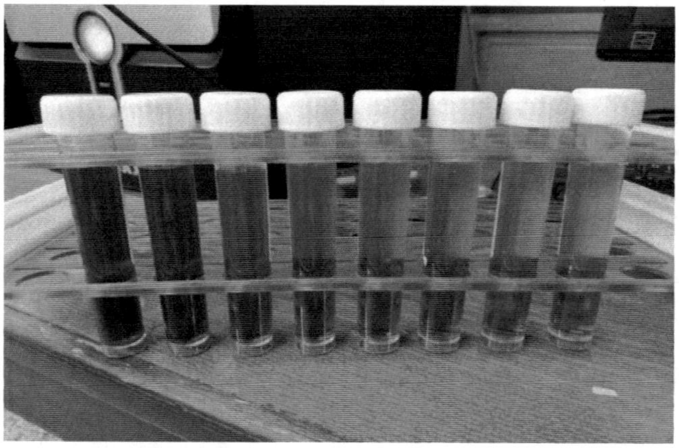

Figure (5) Deceleration of Indigo Carmine.

4. Reaction Kinetic Study of Indigo Carmine

The degradation of indigo carmine can be expressed by the relation between $OH^{\bullet}$radical and concentration of dye:

$$-\frac{dc}{dt} = K.\ [C_{I.C.}]^{n}.\ [OH^{\bullet}]^{m} \quad (6)$$

Working in a galvanostatic environment, a concentration of $OH^{\bullet}$ tendency to be constant, thus $K.\ [OH^{\bullet}]^{m}$ can be written as $K^{*}.\ C_o{}^{q}$:

$$K.\ [OH^{\bullet}]^{m} = K^{*}.\ C_o{}^{q} = K^{**} \quad (7)$$

So eq. (6) can be simplified to:

$$-\frac{dc}{dt} = \acute{o}\ K^{**}.\ [C_{I.C.}]^{n} \quad (8)$$

We can obtain the order of reaction through the relationship between $\ln(C/C_o)$ and time as shown in figure (6).

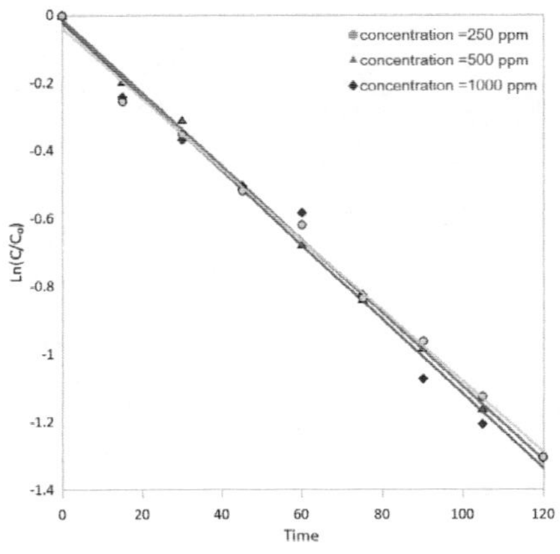

Figure (6) Kinetic fitting curve of Indigo Carmine direct removal (current density 9 mA/ cm^2, at 40°C)

Table (1) the kinetic factors of first-order reactions at different concentrations of Indigo Carmine removal procedure

C_o (mg/l)	K** (min^{-1})	R^2
250	0.0093	0.992
500	0.011	0.9982
1000	0.0096	0.987

we conclude from figure (6) that the relationship is linear, so the reaction became first order and that agrees with searcher Huimin Yang [24].

Integrate eq. (8) with following boundary conditions:

$C = C_o$ at $t = 0$

$C = Ct$ at $t = t$

$$-\int_{C_0}^{C_t} \frac{dC}{C} = K^{**}. \int_0^t t\,.dt$$
(9)

$$-\ln (C - C_o) = K^{**}. t$$
(10)

Taking ln for both sides of eq. (7) yields eq. (11):

$\ln K^{**} = \ln K^{*} + q \ln C_o$
(11)

By plotting $\ln K^{**}$ and as shown in figure (7) where the intercept is equal to $\ln K^{*}$ and the slope is equal to q which is the graphical representation of eq.11

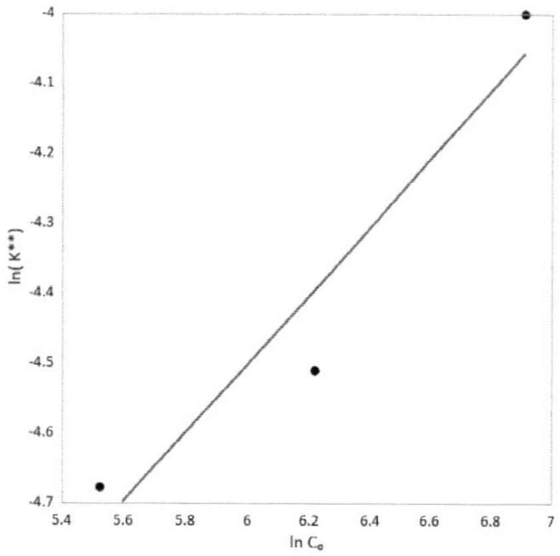

Figure (7) Fitting between $\ln C_o$ and $\ln K^{**}$.

Slop (q) = 0.4887
Intercept ($\ln K^{*}$) = −7.4327, taken exponetional to get value of K^{*}
K^{*}= 5.9125*10^{-4} mg/L.min = 9.8598* 10^{-6}mg/L.s.
The values of the correlation coefficient (R^2) tabled in table (1) implies that equation (11) is acceptable and the relation between $OH^{\bullet}$ conc. and Indigo Carmine conc.is credible.
Equation (12) shows the removal rate of indigo carmine.

$-\frac{dc}{dt} = \; = 9.8598*10^{-6}. \; C_o^{0.4887}. \; C_{I.C.}$
(12)

5. Results and Discussion

5.1 Effect of current density

The concentration (250-1000 mg /L) at 20°C was tested for the effect of current density (3-9 mA/$\mathbf{cm^2}$). The degradation of indigo carmine was noted to increase linearly with current density, so when a high amount of charge enter, the electrolysis will produce more hydroxyl radicals and this explanation agree with Noor H Jawad and Sarmad T Najim [25] [26]

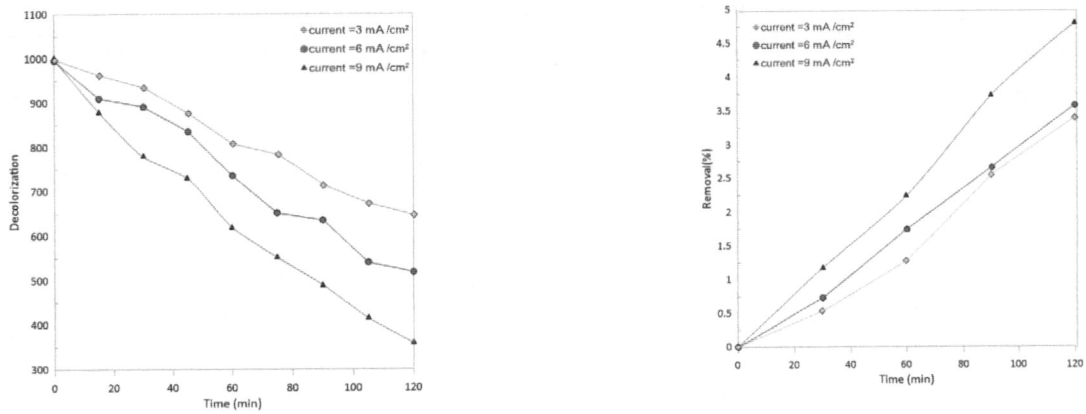

Figure (8) Indigo carmine decolorization and COD removal percent at changing current densities percent with time (1000 ppm, at 20°C)

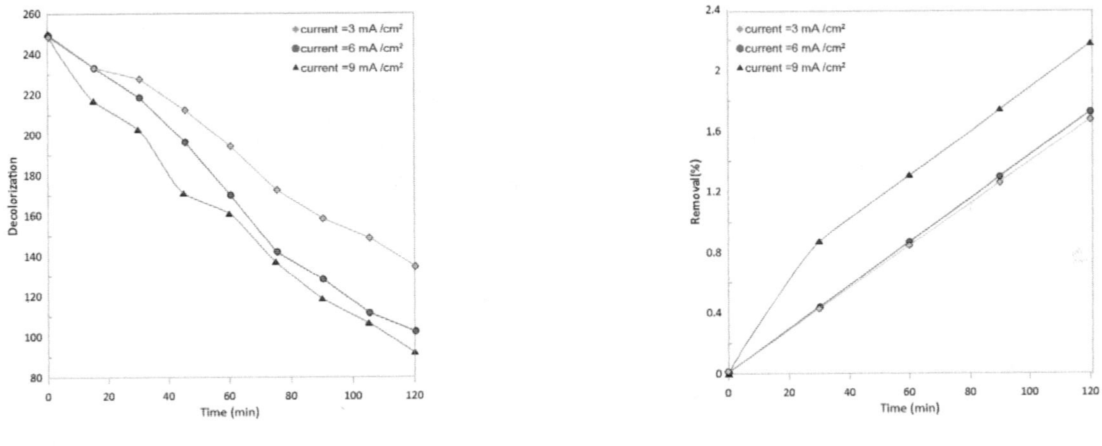

Figure (9) Indigo carmine decolorization and COD removal percent at changing current densities percent with time (250 ppm, at 20°C)

5.2 Effect of electrolyte temperature

Initially, the temperature was controlled by a water bath and a thermometer. The rate of degradation indigo carmine influence by temperature was increasing gradually until it arrives at 40°C then reduce. Generally, chemisorption effect with temperature so this rising in rate shown the chemisorption such as chemical reaction needs active energy. Then when passed 40°C the rate beginning to reduce because the chemisorption process. The desorption of some atoms of the adsorbent chemically bound form above 40°C. And oppositely occurs adsorption to low temperature until 40°C at graphite electrode radicals and this explanation agreement with Shilpa.R and el al [7].

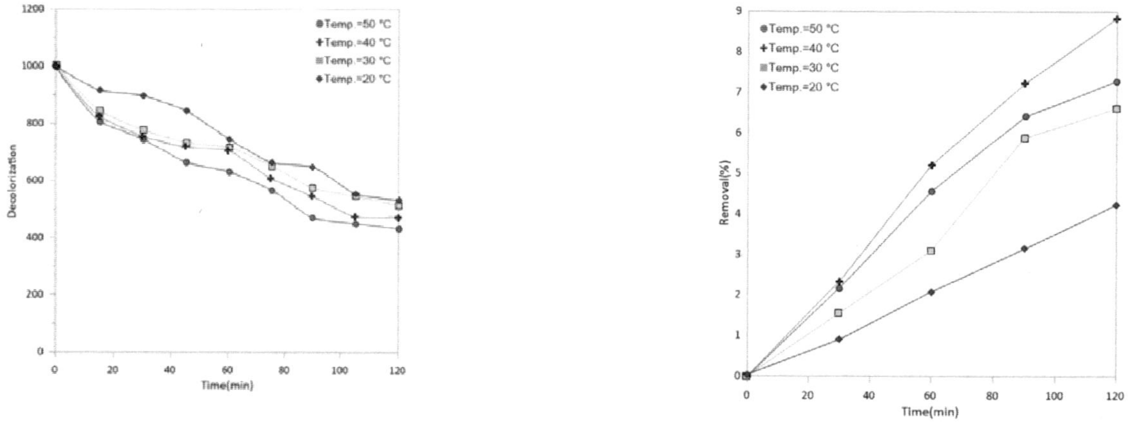

Figure (10) Electrolyte temperature effect on the decolorization and COD removal percent (1000 ppm Indigo carmine at 6 mA /cm^2)

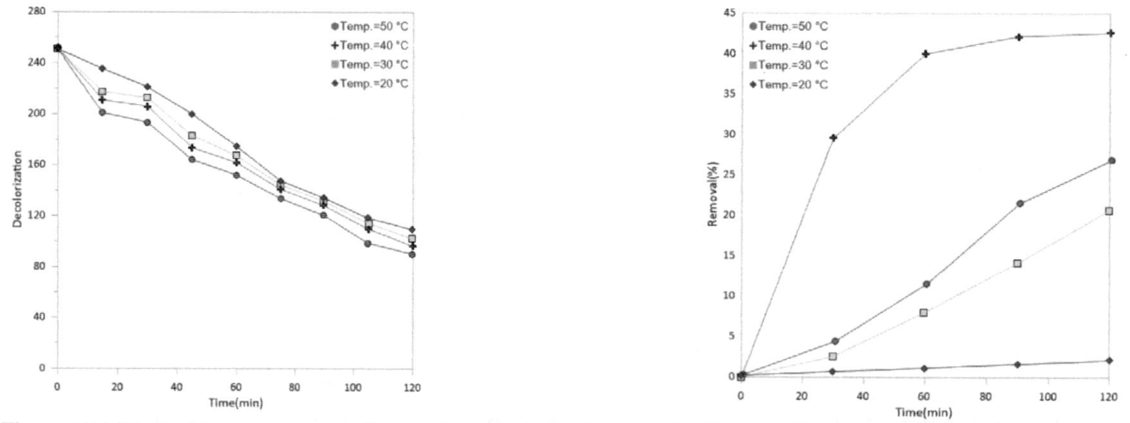

Figure (11) Electrolyte temperature effect on the decolorization and COD removal percent (250 ppm Indigo carmine at 6 mA /cm^2)

5.3 Effect of Initial Dye Concentration

Generally, the degradation of Indigo carmine slightly effected by dye concentration while the rate of degradation will decrease with increasing concentration. this behavior can be explained that the dye will be formed a thin film on the surface of anode so that it reduces the amount of current passed in electrolyte which has an impact on the generation of hydroxyl radical, and this explanation agreement with Charan Kumar H. C and el al[9].

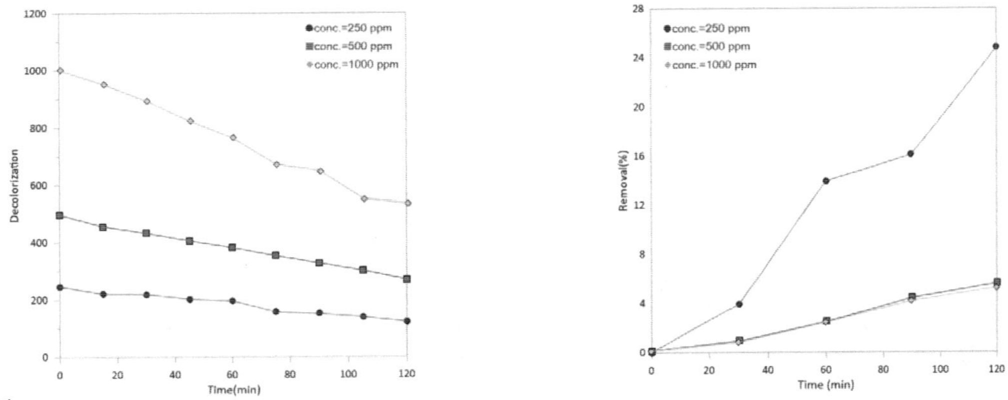

Figure (12) Effect of initial dye concentration on the decolorization and COD removal percent at (30°C and at 3 mA /cm^2).

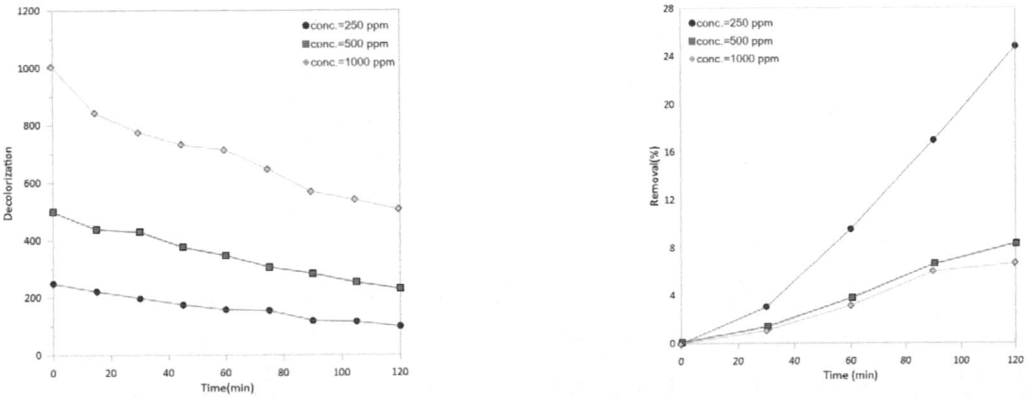

Figure (13) Effect of initial dye concentration on the decolorization and COD removal percent at (30°C and at 6 mA $/\mathrm{cm}^2$)

6. Conclusion

Electrochemical direct oxidation of Indigo carmine used for the industrial textile effluent treatment. (Anode/Cathode) used as graphite electrodes and the trial period was 120 min. The optimum condition is current density (9 mA/cm^2), temperature (40°C,) concentration (250 ppm), electrodes distance (0.5cm), power consumption (3.238 w.h/m^3) , removal percent (61.7%), and found the kinetic reaction first order. The degradation rate increased with rising current density and electrolyte temperature while the rate decreased with rising initial concentration of dye.

7. Reference

1. Bhagawan D., Chandan V., Srilatha K., Shankaraiah G., Rani,& Himabindu V. (2018). Industrial wastewater treatment using electrochemical process. IOP Conference Series: Earth and Environmental Science, 191(1). https://doi.org/10.1088/1755-1315/191/1/012022

2. Abdel-Fatah, Shaarawy H.,Hawash S. (2019). Integrated treatment of municipal wastewater using advanced electro-membrane filtration system. *SN Applied Sciences*, *1*(10), 1–8. https://doi.org/10.1007/s42452-019-1178-9

3. Garcia-Segura S,Ocon J,Chong M. (2018). Electrochemical oxidation remediation of real wastewater effluents — A review. *Process Safety and Environmental Protection*, *113*, 48–67. https://doi.org/10.1016/j.psep.2017.09.014

4. Chequer F. Oliveira G.,and et al.(2013). *Textile Dyes : Dyeing Process and Environmental Impact*.

5. Afanga H,Zazou H,and Titchou F. (2020). Integrated electrochemical processes for textile industry wastewater treatment: system performances and sludge settling characteristics. *Sustainable Environment Research*, *30*(1), 1–11.

6. Stergiopoulos D., Dermentzis K., Giannakoudakis P. and Sotiropoulos S. (2014). Electrochemical decolorization and removal of indigo carmine textile dye from wastewater. *Global Nest Journal*, *16*(3 SPEC. ISSUE), 499–506. https://doi.org/10.30955/gnj.001330

7. Shilpa R, Charan Kumar H.C, Ravishankar Rai.V and Ananda.S. (2017). Electrochemical Degradation of indigo carmine Dye at Pd/graphite Modified Electrode in Aqueous Solution Shilpa.R,. *IOSR Journal of Applied Chemistry*, *10*(12), 47–61. https://doi.org/10.9790/5736-1007030110

8. Wei Lia, Bingnan Mua, Yiqi Yang. (2019). Feasibility of industrial-scale treatment of dye wastewater via bio-adsorption technology. *Bioresource Technology*, *277*(December 2018), 157–170. https://doi.org/10.1016/j.biortech.2019.01.002

9. Charan Kumar H. C, Shilpa. R, RaviShankar Rai. V and Ananda. S. (2017). Electrochemical Degradation of indigo

carmine Dye at Ru / graphite Modified Electrode in Aqueous Solution. *IOSR Journal of Applied Chemistry*, *10*(12), 47–61. https://doi.org/10.9790/5736-1012014761

10. Ammar S., Abdelhedi R., Flox C.,and et al. (2006). Electrochemical degradation of the dye indigo carmine at boron-doped diamond anode for wastewaters remediation. *Environmental Chemistry Letters*, *4*(4), 229–233. https://doi.org/10.1007/s10311-006-0053-2

11. Benkhaya S., Souad M., & Harfi A. El. (2020). A review on classifications , recent synthesis and applications of textile dyes. *Inorganic Chemistry Communications*, *115*(March), 107891. https://doi.org/10.1016/j.inoche.2020.107891

12. Ibrahim M. H., & Najim S. T. (2017). Determination of Mass Transfer Coefficient for Copper Electrodeposition by Limiting Current Technique. *Al-Nahrain Journal for Engineering Sciences (NJES)*, *20*(3), 666–672.

13. Liu Y. X., Liao Z. Y., Wu X. Y., Zhao C. J., Lei Y. X., & Ji D. Bin. (2015). Electrochemical degradation of methylene blue using electrodes of stainless steel net coated with single-walled carbon nanotubes. *Desalination and Water Treatment*, *54*(10), 2757–2764. https://doi.org/10.1080/19443994.2014.903524

14. El-Ashtoukhy E. (2013). Removal of indigo carmine dye from synthetic wastewater by electrochemical oxidation in a new cell with horizontally oriented electrodes. *International Journal of Electrochemical Science*, *8*(1), 846–858.

15. Chatzisymeon E., Xekoukoulotakis N. P., Coz A., Kalogerakis N., & Mantzavinos D. (2006). Electrochemical treatment of textile dyes and dyehouse effluents. *Journal of Hazardous Materials*, *137*(2), 998–1007. https://doi.org/10.1016/j.jhazmat.2006.03.032

16. Ardhianto R., & Bagastyo A. Y. (2019). Personal Care Wastewater Treatment with Electro-coagulation and Electro-oxidation. In *E3S Web of Conferences* (Vol. 125). https://doi.org/10.1051/e3sconf/201912503008

17. Muddemann T., Haupt D., Sievers M., & Kunz U. (2019). Electrochemical Reactors for Wastewater Treatment. *ChemBioEng Reviews*, *6*(5), 142–156. https://doi.org/10.1002/cben.201900021

18. Nurhayati E. (2012). A Brief Review on Electro-generated Hydroxyl Radical for Organic Wastewater Mineralization. *Jurnal Sains &Teknologi Lingkungan*, *4*(1), 24–31. https://doi.org/10.20885/jstl.vol4.iss1.art3

19. Özyurt B., & Camcıoğlu Ş. (2018). Applications of Combined Electrocoagulation and Electrooxidation Treatment to Industrial Wastewaters. *Wastewater and Water Quality*. https://doi.org/10.5772/intechopen.75460

20. Martínez-Huitle C. A., & Brillas E. (2009). Decontamination of wastewaters containing synthetic organic dyes by electrochemical methods: A general review. *Applied Catalysis B: Environmental*, *87*(3–4), 105–145. https://doi.org/10.1016/j.apcatb.2008.09.017

21. Panizza M., Barbucci A., Ricotti R., & Cerisola G. (2007). Electrochemical degradation of methylene blue, *54*(May 2006), 382–387. https://doi.org/10.1016/j.seppur.2006.10.010

22. Nidheesh P. V., Kumar A., Syam Babu D., Scaria J., & Suresh Kumar M. (2020). Treatment of mixed industrial wastewater by electrocoagulation and indirect electrochemical oxidation. *Chemosphere*, *251*, 126437.

https://doi.org/10.1016/j.chemosphere.2020.126437

23. Vasudevan S. (2016). Can Electrochemistry Make the Worlds Water Clean? – A Systematic and Comprehensive Overview. *International Journal of Waste Resources*, *6*(2), 1–5. https://doi.org/10.4172/2252-5211.1000210

24. Yang H., Liang J., Zhang L., & Liang Z. (2016). Electrochemical oxidation degradation of methyl orange wastewater by Nb/PbO2 electrode. *International Journal of Electrochemical Science*, *11*(2), 1121–1134.

25. Jawad N. H., & Najim S. T. (2018). Removal of Methylene Blue by Direct Electrochemical Oxidation Method Using a Graphite Anode. In *IOP Conference Series: Materials Science and Engineering* (Vol. 454). https://doi.org/10.1088/1757-899X/454/1/012023

26. Wang G., Liu Y., Ye J., Lin Z., & Yang X. (2020). Electrochemical oxidation of methyl orange by a Magnéli phase Ti4O7 anode. *Chemosphere*, *241*, 125084. https://doi.org/10.1016/j.chemosphere.2019.125084

Cadmium resistant bacteria mediated cadmium removal: a systematic review on resistance, mechanism and bioremediation approaches

Nadia Jebril[1,2*] **, Rich Boden**[2] **and Charlotte Braungardt**[3]
[1]Department of Biology, College of Sciences for Women, University of Babylon, Iraq. [2] School of Biological and Marine Sciences, University of Plymouth, Drake's Circus, Plymouth PL4 8AA, United Kingdom. [3] School of Geography, Earth and Environmental Sciences and Plymouth University, Drake Circus, Plymouth, PL4 8AA, United Kingdom.

*Email: nadia.tawfiq@uobabylon.edu.iq

Abstract. Cadmium-resistant bacteria that are used to remove cadmium (Cd) are becoming increasingly of the most important and hygienic method. Resistant mechanisms are involved in different ways, and some of them which can be used in cadmium removal techniques based on their molecular mechanisms and minimum inhibitory concentration (MIC). This review summarises recent improvements in understanding the mechanisms by which bacteria are either intrinsically resistant or acquire resistance to cadmium to be used as a way for cadmium removal.

Keywords: Bioremediation, cadmium-resistant bacteria, resistant mechanisms, Minimum inhibitory concentration.

Introduction

Some metals are essential for living organisms in trace amounts, such as cobalt, copper, iron, manganese, molybdenum, and vanadium. In contrast, others, such as cadmium, chromium, mercury, lead, arsenic and antimony, are non-essential. Both essential and non-essential metals cause toxic effects, depending on the dose an organism is experiencing (Bernard, 2008) and its speciation and availability in environments (Jebril *et al*., 2021a). Cadmium is a non-essential element for life, and it is toxic to *Homo Sapiens*. The United States Environmental Protection Agency (US EPA) and the International Agency for Research on Cancer (IARC) classified some elements, including Cd, As, Hg, and Pb, as human carcinogens (ATSDR, 2012). Furthermore, chronic exposure to low levels of Cd is associated with several diseases, such as deranged blood pressure regulation, osteoporosis, early onset of diabetic renal complications, and end-stage renal failure. Humans may be exposed to Cd by drinking polluted water, ingesting contaminated food, and inhaling polluted air, which causes diseases. For example, end-stage renal disease was reported in a Swedish population, living near a Cd battery plant, and using renal replacement therapy (Hellström *et al*., 2001). Skin exposure to Cd is rare, and it occurs mainly as an occupational hazard through contact with Cd-contaminated workplaces (ATSDR *et al*., 2012). After absorption, the blood transfers Cd into different organs, such as the liver, kidney, testis, lungs, heart, prostate, and bone, potentially with harmful consequences. One of the biggest concerns about the presence and accumulation of Cd in the human organs is that the Cd can persist for many years, as it has a long biological half-life of 30 years (ATSDR *et al*., 2012). Standard and guidelines for Cd in various media include (ATSDR *et al*., 2012): Soils with concentrations of > 1 mg/kg Cd are considered contaminated with Cd; Drinking water standard guideline concentration is 3 µg/L (WHO, 2011); Daily intake through food of Cd is 7 µg Cd/kg body; Air pollution exposure is > 5 ng/mL. The mechanism of Cd toxicity is associated with reactive oxygen species (ROS), despite the Cd not being observed as a Fenton metal. Cd is capable of protein depletion and glutathione-bound sulfhydryl groups, resulting in an oxidant, such as the hydroxyl radical species, in cellular environments. These ROS lead to the peroxidation of the lipid to evolve until the DNA is damaged (Waisberg *et al*., 2003). Animal, molecular, and epidemiological models have shown that exposure to Cd causes remarkable health effects upon humans. Interestingly, the depletion of glutathione, which is the dominant form of Cd-induced hepatic toxicity, was observed in the liver of a rat (Liu *et al.*, 2009). Cd toxicity is usually determined by measuring its concentration and its biologically effective dose at the target organ. A study on Australian people between the ages of 41 to 50 years, who died of accidental causes, showed Cd concentrations in the kidney cortex and liver were 26 and 1 μg Cd/g wet weight, respectively (Satarug *et al.*, 2002). The use of a biomarker for exposure is the best way to determine the biologically effective dose when it is impossible to decide the dosage of Cd at the target tissue. Commonly, the biomarker of exposure to Cd is determined by quantifying Cd levels in urine or blood. The level of Cd in urine reflects the accumulation of Cd in the body or, more commonly, the kidney load of Cd, while the level of Cd in blood indicates possible recent exposure to Cd (Wittman, 2002). Most epidemiological studies for Cd exposure in humans showed that the main target organs for Cd toxicity are the liver and the kidney, due to having a high concentration of metallothionein protein (Waalkes and Klaassen, 1985). Biomarker studies for assessing Cd-induced kidney damage in the general population have provided indications of the association between Cd and critical renal impairment concentrations between 100 and 200 μg/g Cd, which links to a urinary threshold limit of 5 – 10 μg Cd/g creatinine (Satarug *et al.*, 2009). The Food and Agriculture Organization/World Health Organization (FAO/WHO, 2010) guideline for a provisional tolerable daily intake of Cd is 1 μg Cd/kg body. Several studies have shown differing kidney effects at urinary Cd levels below the urinary threshold limit of 5-10 μg Cd/g creatinine. The elevation of protein markers, beta-2 macroglobulin, albumin, and retinol-binding protein, in a population with urinary Cd contamination, living in an area near a zinc smelter, have presented with > 2 μg Cd/g creatinine (Trzcinka-Ochocka *et al.*, 2004). Women living in an area contaminated with low Cd also revealed a positive association between Cd concentration and renal damage, which showed that the protein marker at a lower concentration (0.67 μg Cd/g creatinine, Åkesson *et al.*, 2005). Several studies have reported an increase in lung cancer (increased excretion of creatinine) from chronic Cd exposure (0.01 μg $Cd/m^3 \geq 1$ year). A very small study in a silver soldering company reported that 90% of ten workers who were exposed to Cd contaminated air from 0.006 to 0.015 mg/m^3, were found to have urinary Cd contamination, 63.0 μg/g creatinine, having urine protein (beta coefficient 10.27) (Choi *et al.*, 2018). Therefore, due to the high association between Cd intake and carcinogenic effects, the standard guideline concentration for Cd in drinking water is 3 μg/L (WHO, 2011). The ecotoxicity of Cd has been established for different organisms. Pavlaki *et al.* (2016) studied acute toxicity (lethal concentration, LC_{50}) in relation to different marine species. *Daphnea magna*, *Lemna minor,* and fish eggs (*Solea seegalensis*) were found to be most sensitive to Cd, for a day with LC_{50} values around 0.03 mg Cd/L (Clément and Lamonica, 2018; Pavlaki *et al.* 2016; Santos *et al.*, 2018). The cyanobacteria *Synechocystis* sp. PCC6803 is more tolerant to Cd for 5 h with a mean LC_{50} of 8.2 mg Cd/L (Du *et al.*, 2019). It is well-known that Cd inhibits soil microalga, for example, *Chlorococcum* sp. MM11 showed a mean LC_{50} of 2.85 mg Cd/L for 4 days (Subashchandrabose *et al.*, 2015). More Cd tolerant organisms include the earthworm, *Eisenia fetida*, for seven days with a mean LC_{50} value of 129.4 (91.6-202.8) mg Cd/L (Du *et al.*, 2014). Because of the toxicity of cadmium, this review summarizing the possible using of bacteria in cadmium removal from environments.

1. Transporters of cadmium to a cell and bacterial toxicity

Although some elements are essential for cellular metabolism in bacteria, high concentrations of metals can have fatal effects on bacteria. Some elements present in nature at low concentrations, such as Cd, are not necessary for bacteria (Schirawski *et al.*, 2002). Nevertheless, organisms, including bacteria, have mechanisms to take up Cd into their cells (Palmer and Skaar, 2016). Mainly, there are two main carriers categories of Cd in a cell: Firstly, the proteins, which are involved in the absorption of essential micronutrients, such as Zn^{2+} and Mn^{2+}, can also transfer Cd across the cell wall. Cd is the structural equivalent of Zn^{2+} and Mn^{2+} and is handled by Zn^{2+} or Mn^{2+} carriers in most bacteria (Palmer and Skaar, 2016). Two transmitters, the inner membrane ATP-ABC (ATP-binding cassette) ZnuACB and ZIP (Zrt, Irt-like Protein, ZRT/IRT), facilitate the absorption of Zn^{2+} in the cell. ZnuACB shows the carrier for Zn^{2+} absorption in *Yersinia pseud otuberculosis* (Figure1A). ZIP carriers are the dominant carrier, which allows Cd to enter a cell (Binet and Poole, 2000). ABC carriers facilitate the transport of Mn^{2+} across the outer membrane. The SitABCD vector (an example of ABC transport protein) is predominantly a Mn^{2+} transporter (Figure 1B) and also facilitates the entrance of Cd into cells (Kehres *et al.*, 2002). Secondly, specific ligands, such as siderophores, located on the cell surface, can transfer Cd to a cell. Siderophore is a protein with low molecular weight and a significant affinity to Fe^{2+}. The proteins FecA, FepA, and FhuA, mediate the absorption of iron, resulting in ferrisiderophore. The protein substrate is ferrous like FhuD, located in the space between the cell wall membranes, which carries a ferrisiderophore complex from the outer layer to the internal membrane (Figure 1C; Palmer and Skaar, 2016). Similarly, the ferrisiderophore appears to contribute to the uptake of Cd^{2+} in the cell (Winkelmann, 2002). After Cd entered the cell, its toxicity appears mainly due to oxidation reactions with biological macromolecules. Fenton`s reaction summarises the oxidative stress of elements, generally, on the bacterial cells and the production of reactive oxygen species (ROS) as shown in the reactions below (Fenton, 1894):

$M^{(n)} + O_2^{\cdot-}\ M^{(n-1)} + O_2$

$2O_2^{\cdot-} + 2H^+\ H_2O_2 + O_2$

$M^{(n-1)} + H_2O_2\ M^{(n)} + HO^{\cdot} + OH^-$

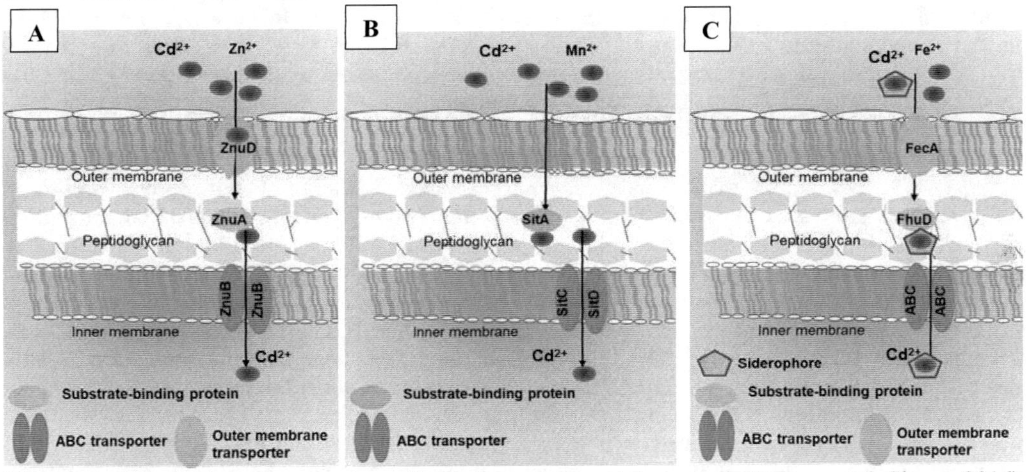

Figure 1. Schematic representation of Cd transposition into a cell (Palmer and Skaar, 2016). **(A)**Transposition of Zn^{2+} in the bacterial cell. Bacteria translocate Zn^{2+} from the extracellular spaces to the cytoplasm by ZnuACB transporters. This Figure is based on the structure of ABC transporter to transport Zn^{2+} into *Yersinia pseudotuberculosis*. **(B)**Transposition of Mn^{2+} in Gram-stain-negative bacteria by SitABCD transporters. **(C)** Fe^{2+} transporter by ABC transporter via siderophore. This structure is based on the structure of ABC transporter to transport Zn^{2+} into *Pseudomonas aeruginosa*.

Hydrogen peroxide (H_2O_2) can be reduced in the presence of element, thus creating one of the most powerful oxidising reagent, the hydroxyl radical ($OH^{\cdot}$). Although Cd does not appear to be working immediately on ROS production, it has been speculated that Cd generates ROS, leading to cell damage (Stohs and Bagchi, 1995). Several other mechanisms enable Cd to damage bacteria (Nies, 2013). The first mechanism is replacing Cd with other ions that occur in functional groups, which are located on the surface of an organism. This substitution is mainly performed with sulfhydryl groups, which are a part of the proteins, or with hydroxyl groups, the main groups of phospholipids, leading to protein damage and lipid oxidation (Chrestensen *et al.*, 2000). Therefore, the toxicity of Cd on the cell is dependent on fatty acids and protein compositions of its cell wall, in addition to the Cd concentration. The second mechanism involves Cd binding with cellular thiols groups, which are part of the proteins, leading to the oxidation of thiols. This oxidation sequentially reduces ATP and NADPH, leading to cell destruction. The third mechanism is DNA damage through $OH^{\cdot}$ oxidation, which occurs when $OH^{\cdot}$ attacks DNA, causing DNA to have permanent reductions in its structure and function. The destruction of the DNA leads to the inhibition of gene expression to proteins and prevents the growth of the cell. The high affinity of Cd with thiol groups is the main chemical basis of Cd toxicity. The superfamily proteins thioredoxin, which has Cys-Gty-Pro-Cys linkage in an active position, such as Dsb (disulfide bond proteins), are placed in the periplasm (Missiakas and Raina, 1997). Most organisms encode mechanisms that protect the proteins from the oxidative damage, and the thioredoxin protein plays an important role in this process (Khairnar *et al.*, 2013). During severe conditions of oxidative stress, these thioredoxin proteins become depleted or overloaded, so the cysteine residues become susceptible to oxidation. The proteins DsbA (disulfide bond A) and DsbB (disulfide bond B) were studied in *E. coli* (Rensing *et al.*, 1997; Stafford *et al.*, 1999) under severe conditions of oxidative stress.

2. Mechanisms of cadmium resistance in bacteria

Some bacteria have a variety of protection mechanisms that increase their tolerance to very high concentrations of Cd without any effect on their growth and metabolism. The protective mechanisms include efflux of the Cd out of the cell, extracellular sequestration, biosorption, precipitation, and bioaccumulation within cells (Bruins *et al.*, 2000, Jebril, 2020a and 2020b). Generally, Cd resistance is defined by the minimum inhibitory concentration (MIC), which is the lowest concentration of a chemical that inhibits the visible growth of a bacterium (Jebril *et al.*, 2022a). The MIC of bacteria is determined by incubating cultures with increasing concentrations of the chemical solution of interest on agar (well-known agar dilution) or in broth (well-known broth dilution) and respectively, evaluating the results by observing the number of bacterial colonies or by determining the biomass of the bacterial growth (Wiegand *et al.*, 2008). Microbes are assumed to be tolerant to Cd if their MIC values exceed 0.8 mM Cd (Matyar *et al.*, 2008). Table 1 illustrates some cadmium strains with their classifications and their natural MIC resistance levels. According to this table, the bacterial phyla *Actinobacteria*, as well as *Firmicute* and *Acidobacteria*, can resist Cd. In addition, strains that belong to Proteobacteria phylum have been reported to have a promising Cd-resistance compared to the other phyla (200 - 700 mM Cd, Ghosh *et al.*, 1997; Mahapatra and Banerjee, 1996). *Proteobacteria* is a major phylum of Gram-stain-negative bacteria, while *Firmicutes* and *Actinobacteria* phyla are the Gram-stain-positive bacteria. Gram-stain-negative bacteria are commonly known to be more resistant to metals compared to Gram-stain-positive bacteria, due to their cell wall composition. Gram-stain-negative bacteria have a compound wall layer, consisting of the outer membrane, the plasma membrane, and the peptidoglycan layer. The outer membrane is a wall through which undesirable elements move from the inside to the outside the cell. The outer membrane consists of an exclusive molecule, LPS (lipopolysaccharide), proteins, lipoproteins, and porins. The plasma membrane, which forms up to 40% of the cell wall, is the high region, separating the outer membrane and the peptidoglycan layer. The peptidoglycan layer is a thin layer of about 3 nm and comprises 10% of the cell wall. In comparison, Gram-stained-positive bacteria contain a single wall layer, consisting of 30 - 90% of the peptidoglycan layer with an estimated thickness of 25 - 30 nm. In addition, secondary polymers, such as teichoic and lipoteichoic acids are formed from cell wall components. Gram-stain-negative bacteria occur in metal-rich environments, and it is reported that, under similar conditions, absorbs less than 10% of the amount of metals that Gram-stain-positive bacteria absorb (Konhauser, 2009). This low absorption rate is due to the metal associated with the functional groups of the outer membrane of Gram-stain-negative bacteria, LPS. Carboxyl groups in the LPS of the cell wall provide binding sites of sufficient electrical charge to bind metal cations to the cell surface effectively. LPS provides enough electrical charge, which can respectively bind with more minerals to the cell surface. LPS consists of three sections: lipid A, a core oligosaccharide, and an O-antigen polysaccharide. Lipid A is the most hydrophobic section, which is composed of glucosamine associated with acyl chains (fatty acids). Chemically, glucosamine and fatty acids are abundant with the carboxylic groups. The second part of LPS is the core, which is linked to the lipid contribution, contains the distinctive sugars of 3-deoxy- *D*-mannooctulosonate and *L*-glycerol-D-mannoheptose, in addition to *N*-acetylglucosamine, galactose and a number of additional sugars, which vary between the bacterial species. These acids and sugars provide large numbers of carboxyl groups, which are the main groups in LPS cores, providing a negative charge for COOH-Cd assembly. The O-antigen polysaccharide is a glycan polymer containing monosaccharides, resulting in a large number of carboxyl groups (Raetz and Whitfield, 2002).

IOP Conf. Series: Earth and Environmental Science 1002 (2022) 012006 doi:10.1088/1755-1315/1002/1/012006

Table 1. Minimum inhibitory concentration of some cadmium resistant strains with their classifications. *No strain code established

Class	Order	Family	Genus and species	Strain code	MIC (mM)	Reference
Phylum: "*Acidobacteria*"						
Acidimicrobiia	*Actinomycetales*	*Actinomycineae*	*Actinomyces turicensis*	AL36Cd	10	Oyetibo *et al*., 2010
Phylum: "*Actinobacteria*"						
Actinobacteria	*Micrococcales*	*Micrococcaceae*	*Arthrobacter ramosus*	AY509238	2.5	Abou-Shanab *et al*., 2007
			A.rhombi	AY509239	5.0	
			Arthrobacter sp.	D9	0.6	Aiking *et al*., 1985
		Microbacteriaceae	*Clavibacter xyli*	AY509235	5.0	Abou Shanab *et al*., 2007
			C.xyli	AY509236	5.0	
			C.xyli	AY509237	5.0	
			Microbacterium arabinogalactanolyticum	AY509224	2.5	

Class	Order	Family	Genus and species	Strain code	MIC (mM)	Reference
			M.arabinogalactanolyticum	AY509226	2.5	
			M.arabinogalactanolyticum	AY509220	0.5	
			M.arabinogalactanolyticum	AY509219	1.0	
			M.arabinogalactanolyticum	AY509221	2.5	
			M.arabinogalactanolyticum	AY509222	5.0	
			M.arabinogalactanolyticum	AY509223	1.0	
	Actinomycetales	*Corynebacteriaceae*	*Corynebacterium ulcerans*	CA56Co	2.0	Oyetibo *et al.*, 2010
			C.kutscheri	FL108Hg	10	
	Actinomycetales	*Streptomyceetaceae*	*Streptomyces*	CdTB01	50	Zhou *et al.*, 2016
Phylum: "*Firmicutes*"						
Bacilli	*Bacillales*	*Bacillaceae*	*Bacillus circulans*	ATCC11778	1.0	Baligarx, 2012
			B. cereus	ATCC11778	1.0	
			B. cereus	S5	10	Wu *et al.*, 2016

IOP Conf. Series: Earth and Environmental Science 1002 (2022) 012006 doi:10.1088/1755-1315/1002/1/012006

Class	Order	Family	Genus and species	Strain code	MIC (mM)	Reference
			B.subtilis	BNi11	2.5	Ndeddy and Babalola, 2017
			B. cereus	BCr26	0.2	
			B.cereus	BNi12	0.1	
			B.cereus	BNi22	2.5	
			B. pumulis	BCd2	5.0	
			B. safensis	BCr7	0.5	
			B.thuringiensis	DM55	0.25	El-Helow *et al.*,2000
			B.cereus	RC-1	0.98	Huang *et al.*, 2013
			B.flexus	AY509229	5.0	Abou-Shanab
			B.niacini	AY509227	0.1	*et al.*, 2007
			B. psychrosaccharolyticus	AY509230	5.0	
			B. megaterium	H3	0.9	Li *et al.*, 2017
		Paenibacillaceae	*Brevibacillus agri*	C15	16 ± 0.7	Jebril *et al.*, 2021b

Class	Order	Family	Genus and species	Strain code	MIC (mM)	Reference
			B. agri	C15 Cd^R	21 ± 0.4	Jebril *et al.*, 2021c
			Geobacillus stearothermophilus	DSM 6453		Hetzer
			G.thermocatenulatus	DSM 5507	0.6	*et al.*, 2006
					0.05	
		Moraxellaceae	*Acinetobacter junni*	CA109Cr	1.0	Oyetibo
						et al., 2010
Phylum "*Proteobacteria*"						
Alphaproteobacteria	*Rhodobacterales*	*Acetobacteraceae*	*Acidiphilium cryptum*	Lhet 2	700	Mahapatra and Banerjee, 1996
			A.symbioticum	KM2	700	
			A.symbioticum	H8	700	
			A.multivorum	JCM	700	
			A.multivoram	GS19h	400	Ghosh
			Acidocella facilis	*	200	*et al.*, 1997

IOP Conf. Series: Earth and Environmental Science 1002 (2022) 012006 doi:10.1088/1755-1315/1002/1/012006

Class	Order	Family	Genus and species	Strain code	MIC (mM)	Reference
	Rhizobiales	*Rhizobiaceae*	*Rhizobium etli*	AY509210	0.1	Abou-Shanab
			R. etli	AY460185	5.0	*et al*., 2007
			R.galegae	AY509213	0.1	
			R. galegae	AY509214	2.5	
			R. galegae	AY509216	0.1	
			R. galegae	AY509215	0.1	
			R. gallicum	AY509211	0.1	
			R.mongolense	AY509212	0.1	
			R.mongolense	AY509209	0.1	
			Sinorhizobium fredii	AY509217	2.5	
	Sphingomonadles	*Sphingomonadaceae*	*Sphingomonas alaskensis*	AY509242	0.1	Abou-Shanab
			S.asaccharolytica			*et al*., 2007
			S.macrogoltabidus	AY509241	0.1	
				AY509243	0.5	

Class	Order	Family	Genus and species	Strain code	MIC (mM)	Reference
	Parvularculaceae	*Caulobacteraceae*	*Caulobacter crescentus*	AY512823	0.1	Abou-Shanab *et al*., 2007
Betaproteobacteria			*Alcaligenes feacalis*	BCd33	7.5	Ndeddy Aka and Babalola , 2017
			A. feacalis	BCr32	2.0	
			A.eutrophus	X58441	10	Abou-Shanab *et al*., 2007
		Comamondaceae	*Acidovorax avenae*	AY512827	2.5	Abou-Shanab
			A.delafieldii	AY512826	0.1	*et al*., 2007
			Variovorax paradoxus	AY512828	5.0	
Gammaproteobacteria	*Pseudomonadales*	*Pseudomonadaceae*	*Pseudomonas putida*	KT2440	0.8	Miller *et al*., 2009
			P.aeruginosa	CA207Ni	10	Oyetibo
			P. aeruginosa	AL80Ni	2.0	*et al*., 2010
			P.aeruginosa	B237	8.0	Baligarx, 2012

IOP Conf. Series: Earth and Environmental Science 1002 (2022) 012006 doi:10.1088/1755-1315/1002/1/012006

Class	Order	Family	Genus and species	Strain code	MIC (mM)	Reference
			P. aeruginosa	BCr3	1.5	Ndeddy and Babalola, 2017
			P.aeruginosa	EP-Cd1	0.45	Muneer *et al*., 2016
			P.riboflavina	AY512822	5.0	Abou-Shanab *et al*., 2007
			P.stutzeri	MTCC101	0.7	Halder and Basu, 2016
			P. aeruginosa	CW-96-1	5.0	Wang *et al*., 1997
Betaproteobacteria	*Burkholderiales*	*Burkholderiaceae*	*Cupriavidus metallidurans*	DSM 2839T	4.5	Jebril *et al*., 2022
	Lysobacterales	*Lysobacteraceae*	*Stenotrophomonas minatitlanensis*	AY512829	0.5	Abou-Shanab *et al*., 2007
	Enterobacteriales	*Enterobacteriaceae*	*Enterobacter cloacae*	CMCB-Cd1	1.90	Haq *et al*., 1999

Class	Order	Family	Genus and species	Strain code	MIC (mM)	Reference
			Escherichia coli	ATCC25922	1.0	Baligarx, 2012
			Klebsiella aerogenes	NCTC 418	0.6	Aiking *et al.*, 1985
			Klebsiella sp.	CMBL-Cd2	0.97	Haq
			Klebsiella sp.	CMBL-Cd3	1.97	*et al.*, 1999
			K. yangling	12	1.51	Hou
						et al., 2015
			K. planticola	Cd-1	15	Sharma
						et al., 2000
			Proteus mirabilis	BNi6	2.5	Ndeddy and Babalola, 2017
			Salmonella enterica	43C	13.3	Khan *et al.*, 2016
Acidothiobacillia	*Acidothiobacilales*	*Acidothiobacillaceae*	*Acidithiobacillus*	DDSM583	500	Baillet *et al.*, 1997

The production of extracellular polymeric substance (EPS) by Gram-satin positive/negative bacteria allows the binding of metal cations to anionic carboxyl groups. The types of EPS vary in terms of the nature of their association with the cells, including C-EPS (capsular), S-EPS (slime), LB-EPS (loosely bound), and TB-EPS (tightly bound) (Żur *et al.*, 2016). EPS consists of carbohydrates, homopolysaccharides, such as either cellulose and dextran or heteropolysaccharides (xanthan, gellan, and alginate) or proteins. Similarly, Poly-*L*-Lysine and poly-gamma-glutamate act as proteins in EPS (More *et al.*, 2014). These carbohydrates and EPS proteins provide sufficient hydroxyl groups for binding Cd, preventing Cd from cytotoxicity and increasing the cell resistance to Cd (Jebril, 2020c, Jebril *et al.* 2021d). Bacteria utilise a range of effective survival mechanisms to assist them in stressful conditions. The mechanisms of bacterial resistance to Cd are shown in Figure 2 A.

(A)

(B)

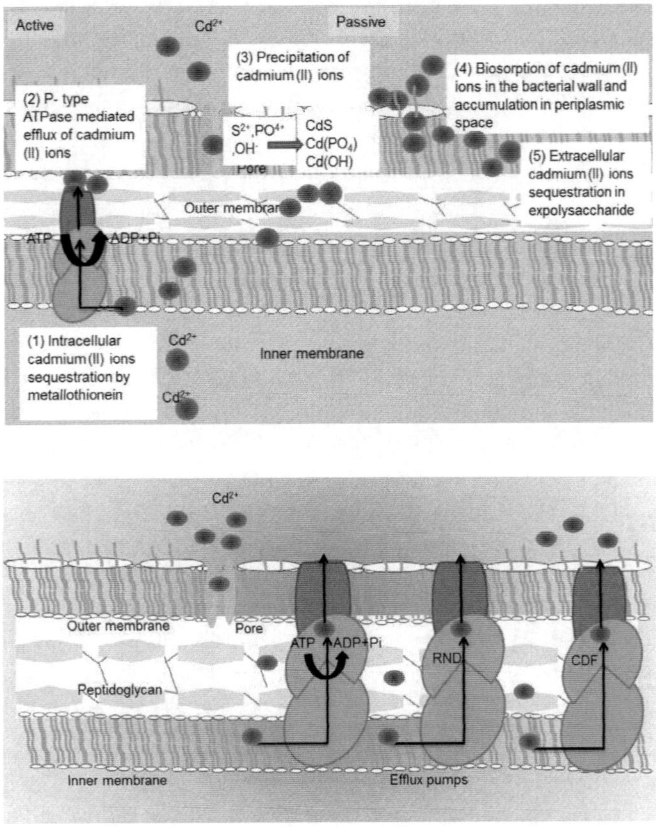

Figure 2. Schematic representation of the bacterial resistance mechanisms to Cd. **(A)** Bacterial defence against Cd. **(1)** Bacteria can deactivate Cd through binding with metallothionein for facilitating the sequestration. (2) The transport of Cd from the cytoplasm to extracellular spaces by efflux pumps. **(3)** Cd can be deactivated via binding with either S, OH, or PO_4 for facilitating precipitation of Cd or **(4)** extracellular substances outside cells for facilitating adsorption of Cd on the bacterial wall (ion exchange, complexation, and physical adsorption). **(5)** Extracellular polymeric substance serves the binding of Cd. The structure was adopted and redrawn according to Prasad *et al.* (2006); Sochor *et al.* (2011); Das and Dash (2017); and Abbas *et al.* (2018). **(B)** Efflux-mediated Cd resistance in bacteria. Cd enters the cell through a transporter. Upon the presence of efflux transporters: P-type ATPase, RND, and CDF, transporters start to pump Cd ions to the out cell. Abbreviations: ATPase (Adenosine triphosphate), RND (resistance-nodulation-cell division) and CDF (cation-diffusion facilitator). The scheme was adopted and redrawn, according to Nies (2003); Rensing and Mitra (2007) and Hynninen (2010).

2.1. Surface biosorption mechanism

P-type ATPase, RND, and CDF, transporters start to pump Cd ions to the out cell. Abbreviations: ATPase (Adenosine triphosphate), RND (resistance-nodulation-cell division) and CDF (cation-diffusion facilitator). The scheme was adopted and redrawn, according to Nies (2003); Rensing and Mitra (2007) and Hynninen (2010). Surface biosorption is an independent metabolism process. Several mechanisms are mediated by surface biosorption such, chelation, complexation, and surface precipitation (Vieira and Volesky, 2000). These mechanisms involve the binding of cadmium to functional groups in the microbial cell wall via electrostatic interactions. As mentioned earlier, the cell walls of the bacteria consist of peptidoglycan, which contains two molecules of NAG (*N*-acetylglucosamine) and NAM (*N*-Acetylmuramic acid), LPS (lipopolysaccharide) and proteins, lipoproteins, porins, and EPS. These components provide several functional groups such as -OH (hydroxyl), -COOH (carboxyl), -PO_3 (phosphate), -NH_2 (ammonia) and -SH (sulfhydryl), which generally have a negative charge, and concurrently a great binding attraction to Cd cation via van der Waals force or covalent bonding (chemisorption). In the adsorption mechanism (Bruins *et al.*, 2000), Cd coordinates with the functional groups, forming cadmium complex, sorbed on the bacterial surface. For example, the reaction of Cd with sulphur bonding is

$Cd + S + 2e^- \rightarrow CdS$ (Xiao *et al.*, 2020).

Following this process of forming CdS, the chelation process could occur (Bruins *et al.*, 2000). As the cells are resistant to Cd exposure, they have evolved their chelation mechanism, such as the extracellular production of substances that can complex with Cd, and eventually, cover its occurrence. The chelation (also well-known as complexation) is a firm binding of Cd with a ligand such as EPS, resulting in a stable cadmium complex. The cadmium chelation is the formation of the coordinate bonds, and it is distinguished from a ligand, which forms a complex compound between the ligand and a single central atom of cadmium. EPSs in bacterial cells are related to their roles in the initial attachment of a cell to different unwanted molecules and Cd biosorption. The carboxyl groups of EPS contribute significantly to the Cd binding site at the bacterial wall of Gram-stain-positive bacteria; whereas, the phosphate groups, which are in the form phospholipids, contribute to Gram-stain-negative bacteria. Recently CadR binding protein, which is responsible for the biosorption of Cd, was found in *E.coli* BL21 (DE3), and the *capB* gene, responsible for Cd adsorption, was present in this strain (Qin *et al.*, 2017).

2.2. Precipitation of cadmium on the bacterial cell surface

The positive charge at cell surfaces resulting from protonation of functional groups at low pH leads to a limitation of Cd cation sorption and their ensuing precipitation on the cell surface, and cadmium remains in solution. Conversely, at a high pH, the functional groups of the cell walls are associated with anions, permitting the approach of Cd cation, resulting in an adequate bond force for sorption to take place. Cell surface precipitation independent of metabolism involves the production of certain compounds when the cells are exposed to Cd, resulting in Cd precipitation. Bacterial cells produce a variety of compounds that react with Cd ions, forming cadmium compounds with different properties. Some of these cadmium compounds have low solubility constants. Some of the theoretical solubility constant in the water at 25 °C are 2.53 $X10^{-33}$ for $Cd_3(PO_4)_2$, 1 $X10^{-28}$ for CdS, 7.2 $X10^{-15}$ for $Cd(OH)_2$, 1.0 $X10^{-12}$ for $CdCO_3$, 2.6 $X10^{-4}$ for CdO and 4.3 $X10^{-1}$ for $CdSO_4$ (Rumble, 2018). With reduced solubility of Cd, the bioavailability and toxicity of the metal are reduced, and precipitation occurs. Macaskie and Dean (1982) studied the first biological precipitation of Cd by *Citrobacteria* sp. with phosphate, which was mediated by phosphatase to accumulate Cd as $CdHPO_4$. Whereas later, Cunningham and Lundie (1993) suggested the precipitation of Cd in *Clostridium thernoaceticum* as

CdS, which was formed by active metabolism of the cells that were able to produce sulfide in a growth culture containing EDTA. It has been reported that sulfides can be produced biologically by several mechanisms, for example, the engineered *E. coli* produced sulfide by heterologous expression of the thiosulfate reductase gene (Bang *et al*., 2000a). The gene responsible for the surface precipitation of Cd is the thiosulfate reductase gene (*phsABC*), which represents the production of hydrogen sulphide, in *Salmonella enterica* (Bang *et al*., 2000b).

2-3. Bioaccumulation of cadmium within the cells

Bioaccumulation is a complex detoxification process that involves either the localisation of metal within specific organelles or the enzymatic transformation of the metal within specific enzymes (Bruins *et al*., 2000). The bioaccumulation of Cd involves the intracellular accumulation of Cd, rather than including the enzymatic pathway, which is the possible transformation for the transmission of metals such as Hg, Se, and Cr. A metal-binding protein, such as the metallothionein can bind with cadmium ions, thereby aiding the accumulation of free-cadmium ions. Metallothioneins are cysteine-rich proteins that are capable of forming stable complexes with different metal cations, such as Cd, Zn, and Cu, to the thiol groups of cysteine (R-SH). Cadmium-metallothioneins sequester Cd within the cell, leading to bioaccumulation and facilitating Cd detoxification by prevention the interaction of Cd with other crucial proteins of cellular metabolism. The first report of bacterial cadmium metallothionein (in *Pseudomonas putida*) was in the early 1980s (Higham *et al*., 1984), and metallothionein was found to be responsible for the sequestration of Cd in *Synechococcus* spp (Li *et al*., 2020). The bioaccumulation of Cd in *Arthrobacter* sp. and *Pseudomonas* sp. has been documented by Scott and Palmer (1990) and Velásquez and Dussan (2009) reported that more ions accumulated in living cells than dead ones, due to two processes occurring: (1) the adhering of Cd onto the cell surface and (2) the accumulation of Cd inside the cell. Heterologous protein, LamB, which was obtained from gene cloning of *E. coli*, enables binding with Cd (Kotrba, 1999). Proteomic approaches showed that SBP`s (selenium binding protein) overexpression protects *Arabidopsis* from the toxic effects of Cd (Hugouvieux *et al*., 2009).

2.4. Efflux pumps

The transport of Cd from the outer membrane to the cytoplasm of the cells is a dynamic process. The transporters (either proteins or siderophores) are present at the cell surface, transferring the Cd and concentrating it within the cell. The basic strategy for reducing Cd concentrations within the cell is to pump Cd outside the cell (Wood and Wang, 1983). The efflux pumps are the central detoxification systems of divalent metal ions, including Cd. The P-type ATPase transporter is the primary efflux pump transporter, which is required for Cd resistant microbes to export Cd from their cytoplasm to periplasm. P-type ATPase transporter belongs to the transmembrane transporter family, the superfamily 1B-2 subgroup (Tsai and Linet, 1993). Three cadmium resistant transporters for the efflux pumps are shown in Figure 2 B: the resistance-nodulation-cell division (RND), the cation-diffusion facilitator (CDF), and the ATP-binding cassettes (ABC), (Leedjärv *et al*., 2008). These transporters can be divided further into two classes, CzcCBA, and Czn systems, which are categorised depending on the mechanism they use to pump Cd out of the cell (Stähler *et al*., 2006). Cd transport in Gram-stain-positive bacteria is similar to those of the Gram-negative bacteria, but their transport functions vary. The cation-proton antiporter in the Gram-stain-negative bacteria CDF is a transporter rather than a cation-transporting ATPase, which is the primary transporter in Gram-stain-positive bacteria (Nies, 2003). The proteins CDF, DmeF, and FieF in *Wautersia metallidurans* CH34 (Munkelt *et al*., 2004), and FrnE protein in *Deinococcus radiodurans* (Khairnar *et al*., 2013) were found to be the efflux proteins for Cd. RND proteins are the leading Cd exporters in Gram-stain-negative bacteria, as it has been found in *Ralstonia metallidurans* CH34 (Nies, 2003).

3. Molecular mechanisms of cadmium regulation in bacteria

Gesturing proteins, transcriptional regulators, and metabolites are the main mediators of the cell`s response to cadmium. Glutathione, glutamyl- cysteinyl-glycine (GSH) is the significant ligand of Cd ions in *E. coli* K-12. GSH contributed to gain resistance against Cd by using a GSH biosynthetic pathway (Suzuki *et al.*, 2005). The activation of methanogenesis studied in *Methanosarcina acetivorans* under cadmium stress found an alternative process that increases ATP yield, which can be used as a biofuel (Lira-Silva *et al.*, 2012). Advancing proteomic analysis has identified the essential proteins to understand the mechanisms that defend microbes against Cd stress. For example, some bacteria resist Cd have been thoroughly studied, and their related genes have been identified. Zhai *et al.* (2017) demonstrated that *L. plantarum* CCFM8610 had prophage P2b 18, CadA, mntA, and lp –3327 proteins, as investigated by the proteomic analysis, and these proteins were associated with the genes: *cadA* and *cadC*. Some proteins have been identified as regulating active Cd translocation between cytoplasm and periplasm of the bacterial cells. These proteins mediate the transporter systems and regulate the Cd efflux, such as *OxyR*, and *OhrR* regulons in *Xanthomonas campestris* (Banjerdkij *et al.*, 2005), and *czrA* regulon in *Caulobacter crescentus* (Valencia *et al.*, 2013). Also, other types of proteins have been identified as regulating the active Cd uptake amongst the members of ZIP transporter, which are mainly involved in the Cd translocation across the cellular membranes. In *E. coli* GG48, two genes (*zntA* and *zitB*) were recognised to encode for ZIP transporter (Grass *et al.*, 2002). Some studies indicate the ability of ZIP transporters to carry and uptake Cd, for example, the transporters ZntA, and RcnA contribute to the Cd uptake of *E. coli* ECA580 (Taudte and Grass, 2010). It has been demonstrated that the *zntA* gene of *E. coli* K-12, encodes the Cd transporter, P-type ATPase ZntA, which belongs to the MerR transcriptional regulator family (Binet and Poole, 2000). Many genes such as *cadA*, *cadB cadC*, *cadD* and *cadX*, and many operons such as *cadA* and *cadC* mediate the efflux system (Busenlehner *et al.*, 2002). The *cadA* operon has been reported in many Cd resistant bacteria such as: *S. aureus* (Nucifora *et al.*, 1989), *Bacillus subtilis* (Tsai *et al.*, 1992), *Listeria monocytogenes* (Lebrun *et al.*, 1994), *Helicobacter pylori* (Herrmann *et al.*, 1999), *Stenotrophomonas maltophilia* (Alonso *et* al., 2000) and *Pseudomonas putida* (Lee *et al.*, 2001). The cadmium operon consists of two genes, the *cadA* gene, which acts as a P-type ATPase transporter, and the *cadC* gene, which acts as a control gene for the *cadA* gene. Figure 3 shows the *cadA* operon and regulation of the expression of Cd resistance in the *S. aureus* pl258 (Nucifora *et al.*, 1989). When *S. aureus* pl258 is exposed to low concentrations of Cd, the Cd occupancy decreases it's binding to the regulatory binding protein of the *Rcad* promoter region, which inhibits the translation of mRNA. While, when *S. aureus* pl258 is exposed to a high concentration of Cd, the high occupation of Cd increases the binding of Cd to the *Rcad* promoter region, allowing the translation of mRNA and the expression of *cadA* gene, which confers the resistance to Cd. The genes, which are responsible for Cd resistance in bacteria, can be located on either plasmids or chromosomes. It is reported that the plasmids can encode a reduced susceptibility to any pollutant, which was firstly involved in the toxic metals (Bruins *et al.*, 2001). However, it has been suggested that the efflux system, which avoids any internal accumulation of Cd, is due to the plasmid-dependent cadmium resistance. The plasmid-dependent cadmium resistance, the operons *cadA*, and *cadB* have been described in *S. aureus* (Kuroda *et al.*, 2001).

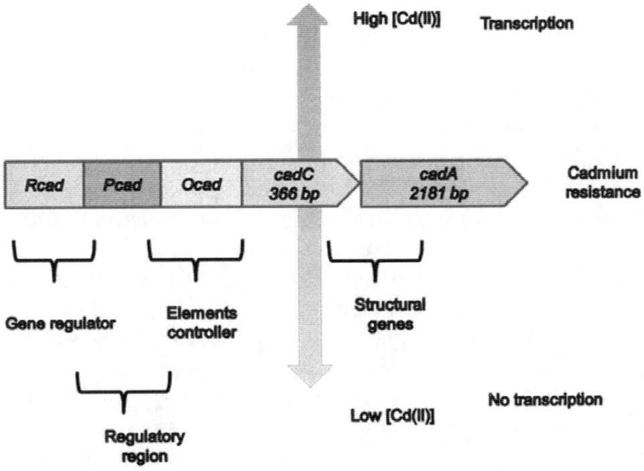

Figure 3. Molecular regulating of cadmium detoxification systems as an example of *cadA* operon structure in *Staphylococcus aureus* pl258 (Nucifora *et al.*, 1989). This model was assumed and redrawn according to the regulation model of the expression of cadmium resistance in *Streptococcus thermophilus* 4134 (Schirawski *et al.*, 2002). The regulatory genes and levels of Cd can relieve mRNA-transcription and control the gene expression of *cadA*. The non-existence of binding protein with Cd at low Cd concentrations causes the *Rcad* promoter region to lose the ability of mRNA transcription. Stimulation of the binding of Cd to the *Rcad* promoter region at high Cd concentration causes the *Rcad* promoter region to gain the ability of the mRNA transcription and increases the gene expression of *cadA*.

In addition, the plasmid-dependent cadmium resistance of the transporter zinc-cobalt, *czcNICBADRS* has been demonstrated in plasmid pMOL30 of *R. metallidurans* CH3 (Nies, 1995). *czcNICBADRS* is an inducible non-ATPase cation efflux Cd, Zn, and Co system. However, the essential Cd resistance systems are usually chromosome-based and are more complex than plasmid-based resistance systems (Khan *et al.*, 2015). Chromosomal mediation of the bacterial resistance to Cd could occur in the gene, gaining the genetic elements overexpression of the cadmium detoxification systems (Bruins *et al.*, 2001). The chromosome of *P. putida* KT2440 includes the genes: *cadA1* and *cadA2* for P-type ATPase and the two CBA transporters are CzcCBA1 and CzcCBA2 (Cánovas *et al.*, 2003). The possibility of the gene expression of the transporters: *cadA1* (P-type ATPase) and *czcCBA1* (CBA) is very low, while the expression of the gene *cadA2* (P-type ATPase) is very high (Leedjärv *et al.*, 2008).

4. Bacterial mechanisms for cadmium bioremediation

The two resistance mechanisms that are capable of achieving cadmium bioremediation are biosorption and bioaccumulation. These mechanisms are physiochemical and/or catabolic processes. Physiochemical processes are non-biological, not dependent on metabolism, passive, and include biosorption (Jebril *et al.*, 2022b). Catabolic processes are biological, active, and defined as the metabolic breakdown of molecules into smaller units, thus strictly dependent on cell metabolism and include bioaccumulation (Kim *et al.*, 2016). However, both metabolic and non-metabolic processes can be responsible for the precipitation of Cd. The bacterial cell wall consists of peptidoglycan, teichoic acids, phospholipids, and lipopolysaccharides, which are the first useful elements for adsorbing Cd due to having anionic functional groups. Therefore, bacterial bioremediation can be divided into three main steps (Bruins *et al.*, 2000): (1) Biosorption is a physicochemical process, in which Cd is passively adsorbed on to the surface of the bacterial cells. The bacteria cell wall consists of peptidoglycan ($C_9H_{17}NO_7$) that has sufficient anionic functional groups that enable binding with Cd. Therefore, biosorption occurs on the surface of either living or non-living bacterial cells, as the functional groups are available in both forms of the bacterial cells. The availability and types of these functional groups control the process, which is also dependent on the species and concentration of Cd, which is determined by the pH and ion strength of the solution. (2) Bacterial cells act as the precipitation surfaces for cadmium, in a process independent of metabolism, resulting in cadmium accumulation. The adherence of Cd on the functional groups of the cell walls through different biosorption mechanisms is followed by the formation of Cd compounds with reagents produced by the cell that has low solubility and hence are prone to precipitation on the cell surface. The potential advantages of the precipitation of Cd as CdS compound over other compounds is the low solubility product compared to hydroxide, carbonate, and phosphate precipitation, leading to a fast precipitation rate and the re-use of sulfide. Several bacterial strains such as *Cupriavidus taiwanensis* KKU2500-3 and *P. aeruginosa* KKU2500-8 (Siripornadulsil and Siripornadulsil, 2013) can precipitate cadmium sulfide and generate sulphide under strict anaerobic or aerobic conditions. These strains can precipitate Cd as CdS due to having a cysteine desulfhydrase that has D-cysteine, which contains hydrogen sulfide (H_2S) (Sakimoto *et al.*, 2016). This dissolved sulfide can then be complexed with Cd to form insoluble CdS precipitates. The precipitation of CdS is typically associated with the bacterial cell wall and occurs predominantly in the periplasmic space. Marusak *et al.* (2016) provided evidence for this process in engineered *E. coli*, from which they harvested, and purified nanoparticles (NP), and NP formed CdS. (3) Following the surface interactions (adsorption and precipitation), Cd-bioaccumulation as an intracellular interaction of cadmium ions with the cell components may occur. As Cd bioaccumulation is more efficient in live cells (Velásquez and Dussan, 2009), using live bacterial cells in bioaccumulation is desirable, and this requires adjustments of nutrients and other environmental conditions during bioremediation.

4.1. Phytoremediation of cadmium

Cd-bioaccumulation is well-applied in the remediation of Cd by rhizobia, diazotrophic bacteria to improve the accumulation of cadmium by plants in phytoremediation (Ike *et al.*, 2007). Phytoremediation has also been used for the treatment of soil contamination with metals. The use of phytoremediation for cadmium and other compounds were first reported in the 1970s (Henry, 2000). Some plants have developed different mechanisms of selective phytoremediation of cadmium from soils. Most Cd phytoremediation in the root and shoot systems occurs via the process called phytoextraction, or through rhizofiltration, the adsorption and precipitation of cadmium in the plant`s roots. Another mechanism of phytoremediation is phytostabilisation, which is the process of reducing the migration of contaminants by limiting their mobility (Kuppusamy *et al.*, 2016).

4.2. Bioremediation of cadmium involving Eextracellular Ppolymeric Ssubstance (EPS)

Bioremediation of Cd may be facilitated by cadmium uptake through the production of EPS. EPS can be used in bioremediation as mixed culture EPS, single-cell EPS, live, dead, or immobilised EPS. EPS contains polysaccharides, proteins, and lipids, which have functional groups such as sulfhydryl, amino, carboxylic, and phosphate groups that enable binding to metals. The accumulation of the metal ions onto the cell wall by EPS is used widely in the bioremediation of Cd and other metals, as listed in Table 2. The removal capacities for metals vary, depending on the reduction potential, atomic weight and the ionic size of the metal. Table 2 shows the variation in the removal capacities for certain metals that are related to the properties of the bacterium, such as active functional groups of EPS. Gupta and Diwan (2017) reviewed the advantages and disadvantages of using EPS in metal bioremediation: EPS is a cost-effective material and can be recycled for the adsorption and the recovery of metal ions. However, EPS shows low abilities for biosorption and requires maintaining the viability of live cells during the biosorption process. In addition to the contribution of EPS of bacterial cells in the detoxification system of Cd, its role in the bioremediation of Cd has been well-studied (Zhou *et al.*, 2009).

Table 2. The removal capacities of some of the bacterial species derived EPS used previously for the bioremediation of some metals.

Bacterial species derived EPS	Removal capacity (nmol /g EPS)	Reference
Gloeocapsa gelatinosa	396 Pb	Raungsomboon *et al.*, 2006
Calothrix marchican	314 Pb	Ruangsomboon *et al.*, 2007
Lactobacillus plantarump	1333 Pb	Feng *et al.*, 2012
Wangia profunda SM-A87	535 Cd	Zhou *et al.*, 2009
Paenibacillus jamilaen	1449 Pb	Morillo *et al.*, 2006
	187 Cd	
Azotobacter chroococcumn	161 Pb	Rasulov *et al.*, 2013
	190 Hg	
Chryseomonas luteola	7.5 Cu	Ozdemir *et al.*, 2005a
	20 Ni	
Chryseomonas luteola	570 Cd	Ozdemir *et al.*, 2005b
Ochrobactrum anthropic	1117 Cr	Ozdemir *et al.*, 2003
	412 Cu	
	263 Cd	

5. Conclusions

Bioremediation of cadmium from water based on natural microbes with resistance to Cd that are isolated in order to augment their resistance then, to further enhance their application in the bioremediation of Cd. Recent developments in the cadmium bioremediation process have increased the need for greater development on three key issues (1) Microbial resistance capability based on microbial relationships and cadmium concentration; (2) Lack of procedural knowledge needed to provide strategic guidance for environmental biotechnology, and (3) Lack of improvement in the bioremediation process.

References

Alonso, A., Sanchez, P., Martínez, J. L. (2000) '*Stenotrophomonas maltophilia* D457R contains a cluster of genes from gram-positive bacteria involved in antibiotic and heavy metal resistance'. *Antimicrob Agents Chemother*, 44:1778-1782.

ATSDR, Agency for Toxic Substances and Disease Registry, Faroon, O., Ashizawa, A., Wright, S., Tucker, P., Jenkins, K., Ingerman, L., Rudisill, C. (2012). 'Toxicological profile for cadmium, '. *Atlanta, Georgia,*

Bang, S.-W., Clark, D. S. and Keasling, J. D. (2000a) 'Engineering hydrogen sulfide production and cadmium removal by expression of the thiosulfate reductase gene

Bang, S.-W., Clark, D. S., Keasling, J. D. (2000b) 'Cadmium, lead, and zinc removal by expression of the thiosulfate reductase gene from *Salmonella typhimurium* in *Escherichia coli'*. *Biotechnol Lett*, 22:1331-1335.

Banjerdkij, P., Vattanaviboon, P., Mongkolsuk, S. (2005). 'Exposure to cadmium elevates expression of genes in the *OxyR* and *OhrR* regulons and induces cross-resistance to peroxide killing treatment in *Xanthomonas campestris'*. *Appl Environ Microbiol*, 71: 1843-2049.

Binet, M. R., Poole, R. K. (2000). 'Cd (II), Pb (II) and Zn (II) ions regulate expression of the metal-transporting P-type ATPase ZntA in *Escherichia coli'*. *FEBS letters*, 473: 67-70.

Bruins, M. R., Kapil, S., Oehme, F. W. (2000). 'Microbial resistance to metals in the environment'. Ecotoxicol Environ Saf, 45:198-207.

Choi, W.-J., Kang, S.-k., Wookyung, C. (2018). Chronic cadmium intoxication with renal injury among workers in a small-sclaed silver soldering company'.[in BMJ Publishing Group Ltd.

Cunningham, D. P., Lundie, L. (1993). 'Precipitation of cadmium by *Clostridium thermoaceticum'*. *Appl Environ Microbiol*, 59:7-14.

Gupta, P., Diwan, B. (2017). 'Bacterial Exopolysaccharide mediated heavy metal removal: a review on biosynthesis, mechanism and remediation strategies'. *Biotechnol Rep*, 13: 58-71.

Chrestensen, C. A., Starke, D. W., Mieyal, J. J. (2000). 'Acute cadmium exposure inactivates thioltransferase (Glutaredoxin), inhibits intracellular reduction of protein-glutathionyl-mixed disulfides, and initiates apoptosis'. *J Biol Chem*, 275: 26556-26565.

Das, S., Dash, H. R. (2017). Handbook of Metal-microbe Interactions and Bioremediation. CRC Press.

Du, Y.-L., He, M.-M., Xu, M., Yan, Z.-G., Zhou, Y.-Y., Guo, G.-L., Nie, J., Wang, L.-Q., Hou, H., Li, F.-S. (2014). 'Interactive effects between earthworms and maize plants on the accumulation and toxicity of soil cadmium'. *Soil Biol Biochem*, 72:193-202.

Du, J., Qiu, B., Gomes, M. P., Juneau, P., Dai, G. (2019) 'Influence of light intensity on cadmium uptake and toxicity in the cyanobacteria *Synechocystis* sp. PCC6803'. *Aquat Toxicol*, 211:163-172.

Hellström, L., Elinder, C.-G., Dahlberg, B., Lundberg, M., Järup, L., Persson, B, Axelson, O. (2001). 'Cadmium exposure and end-stage renal disease'. *Am J Kidney Dis*, 38:1001-1008

Henry, J. R. (2000). An overview of the phytoremediation of lead and mercury. US Environmental Protection Agency, Office of Solid Waste and Emergency Response, Technology Innovation Office Washington, DC.

Ghosh, S., Mahapatra, N., Banerjee, P. (1997). 'Metal resistance in Acidocella strains and plasmid-mediated transfer of this characteristic to Acidiphilium multivorum and *Escherichia coli*'. *Appl Environ Microbiol*, 63: 4523-4527.

Grass, G., Wong, M. D., Rosen, B. P., Smith, R. L., Rensing, C. (2002). 'ZupT is a Zn (II) uptake system in *Escherichia coli'*. *JB*, 184: 864-866.

Jebril N. (2020a). Systems biology design of the microbe toward the bioremediation of cadmium: a review *Int. J. Psychosoc*, 24 (8): 9160- 9166.

Jebril NMT.(2020b). *In Vitro* Bioremediation: A Development Process of Cadmium and Mercury Removal by Environmental Biotechnologies of UV-Mutated *Escherichia coli* K12 and *Bacillus subtilis* 168. *B Sci.J* 17(1(Suppl.):0244-254.

Jebril NMT (2020c) Evaluation of two fixation techniques for direct observation of biofilm formation of *Bacillus subtilis* in situ, on Congo red agar, using scanning electron microscopy, *Vete Wor*, 13(6): 1133-1137.

Jebril, N., Boden, R. & Braungardt, C. (2021a) The effect of pH, calcium, phosphate and humic acid on cadmium availability and speciation in artificial groundwater. *J. Phys.: Conf. Ser.* 1879 022020

Jebril, N., Boden, R. & Braungardt, C. (2021b) 'The isolation and identification of cadmium-resistant *Brevibacillus agri* C15', *J. Phys.: Conf. Ser.* 022015:1-15

Jebril, N., Boden, R. and Braungardt, C. (2021c). The use of ultraviolet-light mutagenesis to generate a mutant with elevated cadmium resistance, *B. agri* C15 Cd^R, *J. Phys.: Conf. Ser.* 022043: 1-20.

Jebril, N., Boden, R. and Braungardt, C. (2021d). Cadmium Removal with Mutant *Brevibacillus Agri* C15 Cd^R Entrapped in Calcium Alginate Gel: a New Process. *Conf. Se.: Earth Environ. Sci.* 761 012028.

Jebril, N., Boden, R. & Braungardt, C. (2022a). Determination of minimal inhibitory concentration of cadmium for *Brevibacillus agri* C15 and *Brevibacillus agri* C15 Cd^R. *AIP Conf Proc*,1-27.

Jebril, N., Boden, R. & Braungardt, C. (2022b). Cadmium removal with mutant Brevibacillus agri C15 CdR entrapped in calcium alginate gel: multi-constituent ionic exchange. *AIP Conf Proc*,1-27.

Khairnar, N. P., Joe, M.-H., Misra, H., Lim, S.-Y., Kim, D.-H. (2013). 'FrnE, a Cadmium-Inducible Protein in *Deinococcus radiodurans*, Is Characterized as a Disulfide Isomerase Chaperone In Vitro and for Its Role in Oxidative Stress Tolerance *In Vivo'*. *J Bacteriol*, 195: 2880-2886.

Kehres, D. G., Janakiraman, A., Slauch, J. M. & Maguire, M. E. (2002). 'SitABCD is the alkaline Mn^{2+} transporter

Kuppusamy, S., Palanisami, T., Megharaj, M., Venkateswarlu, K., Naidu, R. (2016). 'In-situ remediation approaches for the management of contaminated sites: a comprehensive overview', *Rev Environ Contam Toxicol.* 236:1-115.

Kim, Y., Bae, J., Park, H., Suh, J.-K., You, Y.-W., Choi, H. (2016). 'Adsorption dynamics of methyl violet onto granulated mesoporous carbon: Facile synthesis and adsorption kinetics'. *Water Res*, 101:187-194.

Konhauser, K. O. (2009). *Introduction to geomicrobiology.* John Wiley & Sons.

Kuiper, N., Rowell, C., Shomar, B. (2015). 'High levels of molybdenum in Qatar's groundwater and potential impacts'. *J Geochem Explor*, 150:16-24.

Khairnar, N. P., Joe, M.-H., Misra, H., Lim, S.-Y., Kim, D.-H. (2013). 'FrnE, a Cadmium-Inducible Protein in *Deinococcus radiodurans*, Is Characterized as a Disulfide Isomerase Chaperone In Vitro and for Its Role in Oxidative Stress Tolerance *In Vivo'*. *J Bacteriol*, 195: 2880-2886.

Kuroda, M., Ohta, T., Uchiyama, I., Baba, T., Yuzawa, H., Kobayashi, I., Cui, L., Oguchi, A., Aoki, K.-i., Nagai, Y. (2001). 'Whole genome sequencing of meticillin-resistant *Staphylococcus aureus'*. *The Lancet*, 357: 1225-1240.

Feng, M., Chen, X., Li, C., Nurgul, R., Dong, M. (2012). 'Isolation and identification of an Exopolysaccharide-Producing lactic acid bacterium strain from chinese paocai and biosorption of Pb (II) by its exopolysaccharide' *J Food Sci.*, 6: 77-83.

Food and Agriculture Organization/World Health Organization (FAO/WHO). 2010. Seventy-third Meeting, Geneva, 8–17 June 2010. Summary and Conclusions. JECFA/73/SC.

Leedjärv, A., Ivask, A., Virta, M. (2008). 'Interplay of different transporters in the mediation of divalent heavy metal resistance in *Pseudomonas putida* KT2440'. *JB*, 190: 2680-2689.

Lira-Silva, E., Santiago-Martínez, M. G., Hernández-Juárez, V., García-Contreras, R., Moreno-Sánchez, R., Jasso-Chávez, R. (2012). 'Activation of methanogenesis by cadmium in the marine archaeon Methanosarcina acetivorans'. *PloS one*, 7: e48779.

Liu, J., Qu, W., Kadiiska, M. B. (2009). 'Role of oxidative stress in cadmium toxicity and carcinogenesis'. *Toxicol Appl Pharmacol*, 238: 209-214.

Marusak, K., Feng, Y., Eben, C., Payne, S., Cao, Y., You, L., Zauscher, S. (2016). 'Cadmium sulphide quantum dots with tunable electronic properties by bacterial precipitation'. *RSC Advances*, 6: 76158-76166.

Matyar, F., Kaya, A., Dinçer, S. (2008). 'Antibacterial agents and heavy metal resistance in Gram-negative bacteria isolated from seawater, shrimp and sediment in Iskenderun Bay, Turkey'. *Sci Total Environ*, 407: 279-285.

Nies, D. H. (2013). 'RND efflux pumps for metal cations'. Microbial Efflux Pumps: *Curr Res*, 1: 79-121.

Palmer, L. D., Skaar, E. P. (2016). 'Transition metals and virulence in bacteria'. *Annu Rev Genet*, 50:. 67-91.

Prasad, M. N. V., Sajwan, K. S., Naidu, R. (2006). Trace elements in the environment: biogeochemistry, biotechnology, and bioremediation. CRC Press.

Pavlaki, M. D., Araújo, M. J., Cardoso, D. N., Silva, A. R. R., Cruz, A., Mendo, S., Soares, A. M., Calado, R., Loureiro, S. (2016). 'Ecotoxicity and genotoxicity of cadmium in different marine trophic levels'. *Environ Pollut*, 215: 203-212.

Rumble, J. R. (2018) 'Solubility product constants of inorganic salts '. *CRC Handbook of Chemistry and Physics*, 99 th ed. Taylor & Francis Group, 5-208-209.

Raungsomboon, S., Chidthaisong, A., Bunnag, B., Inthorn, D., Harvey, N. W. (2006). 'Production, composition and Pb^{2+} adsorption characteristics of capsular polysaccharides extracted from a cyanobacterium *Gloeocapsa gelatinosa*'. *Water Res*, 40: 3759-3766

Rasulov, B. A., Yili, A., Aisa, H. A. (2013). 'Biosorption of metal ions by exopolysaccharide produced by *Azotobacter chroococcum* XU1'. *J Environ Prot* , 4: 989-993.

Rensing, C., Mitra, B., Rosen, B. P. (1997). 'Insertional inactivation of dsbA produces sensitivity to cadmium and zinc in *Escherichia coli'*. *J Bacteriol*, 179:: 2769-2771.

Rensing C., Mitra B. (2007). Zinc, Cadmium, and Lead Resistance and Homeostasis. *In*: Nies D.H., Silver S. (*eds*) Molecular Microbiology of Heavy Metals. Microbiology Monographs, vol 6. Springer, Berlin, Heidelberg: 321-341.

Ozdemir, G., Ceyhan, N., Manav, E. (2005a). 'Utilization in alginate beads for Cu (II) and Ni (II) adsorption of an exopolysaccharide produced by *Chryseomonas luteola* TEM05'. *World J Microbiol Biotechnol*, 21: 163-167.

Ozdemir, G., Ceyhan, N., Manav, E. (2005b). 'Utilization of an exopolysaccharide produced by *Chryseomonas luteola* TEM05 in alginate beads for adsorption of cadmium and cobalt ions'. *Bioresour. Technol*, 96: 1677-1682.

Ozdemir, G., Ozturk, T., Ceyhan, N., Isler, R., Cosar, T. (2003). 'Heavy metal biosorption by biomass of *Ochrobactrum anthropi* producing exopolysaccharide in activated sludge'. *Bioresour Technol*, 90: 71-74.

Taudte, N., Grass, G. (2010). 'Point mutations change specificity and kinetics of metal uptake by ZupT from *Escherichia coli'*. *Biometals*, 23: 643-656.

Tsai, K.-J., Yoon, K. P., Lynn, A. R. (1992). 'ATP-dependent cadmium transport by the *cadA* cadmium resistance determinant in everted membrane vesicles of *Bacillus subtilis*'. *JB*, 174: 116-121.

Trzcinka-Ochocka, M., Jakubowski, M., Razniewska, G., Halatek, T., Gazewski, A. (2004). 'The effects of environmental cadmium exposure on kidney function: the possible influence of age'. *Environ Res*, 95:143-150.

Satarug, S., Baker, J. R., Reilly, P. E., Moore, M. R., Williams, D. J. (2002). 'Cadmium levels in the lung, liver, kidney cortex, and urine samples from Australians without occupational exposure to metals'. *Arch Environ Health*, 57:69-77.

Satarug, S., Garrett, S. H., Sens, M. A., Sens, D. A. (2009). 'Cadmium, environmental exposure, and health outcomes'. *Environ Health Perspect*, 118: 182-190.

Schirawski, J., Hagens, W., Fitzgerald, G. F., van Sinderen, D. (2002). 'Molecular characterization of cadmium resistance in *Streptococcus thermophilus* strain 4134: an example of lateral gene transfer'. *Appl Environ Microbiol.*, 68: 5508-5516.

Sochor, J., Zitka, O., Hynek, D., Jilkova, E., Krejcova, L., Trnkova, L., Adam, V., Hubalek, J., Kynicky, J., Vrba, R. (2011). 'Bio-sensing of cadmium (II) ions using *Staphylococcus aureus'*. *Sensors*, 11:10638-10663.

Stähler, F. N., Odenbreit, S., Haas, R., Wilrich, J., Van Vliet, A. H., Kusters, J. G., Kist, M., Bereswill, S. (2006). 'The novel Helicobacter pylori CznABC metal efflux pump is required for cadmium, zinc, and nickel resistance, urease modulation, and gastric colonization'. *Infect Immun*, 74: 3845-3852.

Suzuki, H., Koyanagi, T., Izuka, S., Onishi, A., Kumagai, H. (2005). 'The yliA,-B,-C, and-D genes of *Escherichia coli* K-12 encode a novel glutathione importer with an ATP-binding cassette'. *J Bacteriol*, 187: 5861-5867.

Subashchandrabose, S. R., Megharaj, M., Venkateswarlu, K., Naidu, R. (2015). 'Interaction effects of polycyclic aromatic hydrocarbons and heavy metals on a soil microalga, *Chlorococcum* sp. MM11'. *Environ Sci Pollut Res*, 22: 8876-8889.

Vieira, M. A. (2019). 'Determination of total concentration and bioaccessible fraction of metals in infant cereals by MIP OES'. J Food Compost Anal, 77:60-65.

Valencia, E. Y., Braz, V. S., Guzzo, C., Marques, M. V. (2013). 'Two RND proteins involved in heavy metal efflux in Caulobacter crescentus belong to separate clusters within proteobacteria'. *BMC microbiol*, 13: 79-84.

Velásquez, L., Dussan, J. (2009). 'Biosorption and bioaccumulation of heavy metals on dead and living biomass of *Bacillus sphaericus'*. *J Hazard Mater*, 167: 713-716.

Waisberg, M., Joseph, P., Hale, B., Beyersmann, D. (2003). 'Molecular and cellular mechanisms of cadmium carcinogenesis'. *Toxicology*. 192: 95-117.

Waalkes, M. P., Klaassen, C. D. (1985). 'Concentration of metallothionein in major organs of rats after administration of various metals'. *Fundam Appl Toxicol*, 5: 473-477.

Wittman, R, Hu H. (2002). Cadmium exposure and nephropathy in a 28-year-old female metals worker. *Environ Health Perspect*,110:1261-1226.

Winkelmann, G. (2002) 'Microbial siderophore-mediated transport' *Biochem Soc Trans*. 30: 691-6.

Wiegand, I., Hilpert, K., Hancock, R. E. (2008). 'Agar and broth dilution methods to determine the minimal inhibitory concentration (MIC) of antimicrobial substances'. *Nat Protoc*, 3:163-192.

World Health Organization (WHO). (2011). 'Cadmium in Drinking-water Background document for development of WHO Guidelines for Drinking-water Quality.

Xiao, X., Han, X., Wang, L.-G., Long, F., Ma, X.-L., Xu, C.-C., Ma, X.-B., Wang, C.-X. & Liu, Z.-Y. (2020) 'Anaerobically photoreductive degradation by CdS nanocrystal: Biofabrication process and bioelectron-driven reaction coupled with *Shewanella oneidensis* MR-1'. *Biochem Eng J*, 154:107466.

Żur, J., Wojcieszyńska, D., Guzik, U. (2016). 'Metabolic responses of bacterial cells to immobilization'. *Molecules*, 21: 958.

Zhai, Q., Xiao, Y., Zhao, J., Tian, F., Zhang, H., Narbad, A., Chen, W. (2017) 'Identification of key proteins and pathways in cadmium tolerance of *Lactobacillus plantarum* strains by proteomic analysis'. *Sci Rep*, 7:1182-1187.

Zhou, W., Wang, J., Shen, B., Hou, W., Zhang, Y. (2009). 'Biosorption of copper (II) and cadmium (II) by a novel exopolysaccharide secreted from deep-sea mesophilic bacterium'. *Colloids Surf B*, 72: 295-302.

Zhou, G., Yang, H., Zhou, H., Wang, C., Fu, F., Yu, Y., Lu, X., Tian, Y. (2016). 'Complete genome sequence of the *Streptomyces* sp. strain CdTB01, a bacterium tolerant to cadmium'. *J Biotechnol*, 229: 42-43

Solid waste: A study of its concept, management methods, and environmental impacts

Suaad Hadi Hassan Al-Taai[1]

[1]Prof.Dr, College of Education Ibn Rushd for Humanities, University of Baghdad, Baghdad, Iraq

suaad.hadi@ircoedu.uobaghdad.edu.iq

Abstract

The study of the impact of solid waste and its management is one of the important topics that has attracted the attention of researchers and those interested in the study of the environment due to its significant effects on human health, living organisms in general, and the environment. The research has addressed the study of the concept of solid waste and its classification, focusing on domestic, industrial, agricultural, commercial, and other waste. The research also dealt with the study of municipal solid waste, the rate of waste generation, and the factors affecting it. It referred to the physical and chemical composition of solid waste, the arrangement of waste management and its objectives, and the most important stages of waste flow. It discussed the methods of waste collection, sorting, and recycling, the most important human characteristics, and the natural factors affecting them, concerning the negative effects of waste on the environment.

Keywords: water, wells, energy, density, moisture, land, air.

1. Introduction

The spread of solid waste everywhere has a clear negative impact on human health and the environment due to its dangerous effects. Hazardous impacts on human health, society, and the economy have been reported, so a large number of researchers are interested in studying how to collect, sort, recycle and manage them to reduce environmental pollution as much as possible. Solid waste is defined as any material that is thrown away by a person because it is no longer needed and is no longer usable by that person at that time, despite the possibility of benefiting from those materials thrown away in another place and at another time [1,2]. Waste is defined as those materials that are thrown out or generated by living organisms in a natural ecosystem. This system deals with them on the basis that they are resources that are used efficiently and effectively and are reused within a clear cycle [3]. Waste can be defined as resulting from the process of use, production, or consumption and causes various pollution in treatable media that can benefit from or get rid of its danger [3].

Waste is the waste resulting from activities that represent the usual movement of human life from the home and includes all residues resulting from the user processes that the holder disposes of or intends to dispose of as materials that are not of value and worthy of preservation. Although it may have value elsewhere or be subject to other conditions that encourage reuse and recycling [4]. Solid waste is one of the problems that burden local governments in third world countries that follow the old traditional methods of dealing with solid waste and do not use scientific tools in decision-making. At the same time, developed countries have gone a long way in developing plans and strategies that deal with this waste as a source of many useful materials that can be recycled [5]

[1] To whom any correspondence should be addressed.

2. Classification of solid waste

2.1 .Household (residential) waste:

It often depends on the daily activities of human beings. Therefore, the economic level and the rate of income are among the main factors in the disposal of household (residential) waste. Residential waste constitutes the largest proportion of solid waste in urban areas, as it constitutes 55%–65% of solid waste. Food residues constitute the main component of household waste, as well as varying proportions of food waste, paper, glass, plastic, leather, metals, and spent electronic devices [1-4, 6].

2.2 . Industrial waste:

What is meant by "industrial" is that which is generated by various industries and that is thrown into the environment is a solid, liquid, or gaseous state. They are of a continuous and diversified nature with the diversity of industries such as chemical industries, metals and leather industries, food industries, and other by-products that lag behind manufactured materials such as peels of fruits and vegetables [1-4, 6, 7].

2.3. Solid commercial waste:

It is commercial waste that is generated from commercial centers such as offices, restaurants, markets, and hotels. It is very similar to household waste in terms of waste quality, except that it differs in terms of the proportion of components and the amount of waste produced. Commercial waste resulting from office work includes large quantities of paper, warehouse waste, and large quantities of cartons and packaging boxes, while restaurants and markets contain a large proportion of food waste [1-3, 6, 7].

2.4 . Municipal services waste:

It is defined as waste collected by municipal institutions and generated by residential homes, shops, and government institutions such as schools, hospitals, and others. It includes waste from garages, squares, and parks, and waste from street sweeping. The main components of municipal waste are paper, cardboard, and organic materials such as food waste and garden plants, plastics, metals, glass, and textiles [1-3, 6, 7].

2.5. Construction and demolition waste:

It is the materials resulting from the demolition and construction of facilities and buildings. They are inert materials and do not pose a danger to human health. Therefore, it can be used in backfilling, road construction, and others. But if it is thrown randomly, it leads to a distortion of the general view of the area, and if it is placed on agricultural lands, it will lead to the loss of these lands. The rubble resulting from construction, restoration, and demolition operations constitutes the constituents of construction waste. An implicit environmental problem, and to reduce its risks, it is necessary to start first from the project design phase, as materials are selected for construction that allows their use and/or reuses in new projects and the development of objectives and methods for importing materials.

2.6 . Medical waste:

It is defined by the World Health Organization (WHO) as clinical waste resulting from the treatment of patients in research centers and medical facilities. It is classified as hazardous waste and accounts for approximately 10–25% of the total waste based on its source, type, and risk factors associated with its handling, storage, transportation, and disposal. Contaminated waste includes pathogenic agents, sharp equipment, surgical waste, chemicals, and radioactive waste [1-3, 6]. The most prominent waste product of medical hospitals is plastic. This includes paws, syringes, plastic tubes, feeder bags, blood bags, and others. Second, glass: It includes glass containers and bottles used in the preservation of pharmaceutical materials, as well as all-glass instruments used in analytical laboratories. Third, fabric, including cotton, gauze, and all the bandages used in the process of tying wounds and fractures (bandage) and others. Fourth, metals, which include metal needles, cans, staples, scalpels, and others. Fifth, organic materials include parts of the human body organs that result from cutting and amputations, as well as waste products resulting from the birth process, such as the placenta. Sixth, paper and cardboard, including papers and forms for medical examination and treatment, and boxes in which medicines, devices, and medical supplies are packed. Seventh, food waste, including all waste resulting from the process of cooking food, as well as the remains of meals provided to sleeping patients and others [8].

2.7 . Hazardous waste:

It is waste that is usually industrial or medical and has negative effects on human health and the environment due to its chemical, physical, and biological properties. This includes clinical waste from therapeutic activities and waste resulting from the manufacture of any pharmaceutical preparations, medicines, organic solvents, inks, dyes, paints, pesticides, and fertilizers [1]. It can be said that hazardous wastes (solid, liquid, and gaseous) are those that are legally allowed to be treated in waste complexes and treated in classified units approved by local authorities [3].

2.8 . Non-hazardous (benign) waste:

It is a group of materials that do not have serious environmental problems and, at the same time, are easy to dispose of in an environmentally safe manner. They include household, commercial, and non-hazardous factory waste [3].

2.9. Green waste:

It includes plant pruning products, which are important, especially as they can be used to produce a good soil conditioner or cubes used in energy production incinerators [3].

2.10. Agricultural solid waste:

All waste and residues result from all agricultural activities, including plant waste, including plant residues, leaves, etc., and animal waste, including animal secretions and slaughterhouse waste [2,4,7]. Waste from the agricultural sector usually consists of crop residues, animal manure, and unwanted agricultural chemicals. The process of estimating the amount of waste generated by the agricultural sector is further complicated by the fact that large quantities of animal waste are used directly as organic fertilizer for the land. It is estimated that between 20% and 50% of animal waste is used in this way within the European Union. It is the waste resulting from various types of agricultural activity and starts from the initiation and confirmation processes, which are the activities of crop collection [4].

2.11. Street waste:

This waste is represented by the dust accumulated on the sidewalks, leaves of trees, discarded water bottles, and the like, which are collected from the city's streets and alleys. However, sometimes such waste is left, especially the dust, which often takes part of the sidewalk into the street [2,4,7].

2.12. Treatment Plant Waste:

These include wastes resulting from water treatment plants and sewage treatment plants, as well as industrial water treatment plants [7].

3. Municipal solid waste

Municipal solid waste mainly consists of materials that can be recycled, and their incoming percentages vary according to the nature of society and the amount of cultural and economic prosperity of the country, the most important of which are [9]:

1. Organic waste: its percentage ranges between 50 and 65 of the total generated waste, and it is one of the most important sources of liquid materials and seepage water, in addition to methane gas in landfill sites.

2. Metals: Iron, steel, tin, and nonferrous waste like copper and aluminum account for 4–6% of waste generated. After classifying the minerals, each can be recycled according to its type.

3. Cardboard and paper: their proportions are estimated at 9–15 of the quantity of waste generated, and paper recycling activities have become very important and constitute an important source of income for individuals and countries, especially after re-sorting and benefiting from them.

4. Fabric and Industrial Textiles: These are mainly produced in the manufacturing process of clothes in specialized factories and others and can be directed to their final destination according to the classification of these materials.
5. Plastic: There are seven types of it.
6. Glass: There isn't much importance nowadays to glass recycling.

4. Solid Waste Composition

Solid waste differs in its physical and chemical composition. However, these compounds are important and necessary in the study as well as the development of correct programs and solutions in waste management. Solid waste components and compositions are divided into five parts:

4.1 . Physical composition:

The physical components of solid waste are important matters for the selection of wheels and machines, as well as the method of the waste collection because the physical composition plays an important role in determining the processes of sorting and reusing some materials or using others as a source of energy, as shown in table (1) that describes the ideal physical composition of solid waste [1,7].

Table 1. The ideal physical composition of solid waste

The Ingredients	Term%	% Ideal Values
Food Waste	6.26	15
Paper	25.45	40
Cartoon	3.15	4
Plastic	2.8	3
Rags	0.4	2
Rubber	0.2	0.5
Skins	0.2	0.5
Garden Wood	0.20	12
Wooden Pieces	1.4	2
Glass	4.16	8
Cans	2.8	6
Ferrous Metals	0.1	1
Non-Ferrous Metals	1.4	2
Dirt, Ash, Bricks	0.10	4

4.2 .Chemical composition:

However, it is possible to know the extent of the danger of these wastes and how to benefit from them in the processes of extracting electrical energy, extracting organic fertilizers, or recycling them [1].

4.3. Moisture content:

Moisture content is expressed as the percentage of moisture content to the weight of the wet or dry material. It has been reported that the amount of moisture content of the waste increases in the winter season due to rainfall and low temperatures [7].

4.4. Density:

The amount of density is important as it determines the mass and volume to be processed [7].

5. Waste generation rate and affecting factors :

The rate of waste generation is an important issue in solid waste management. It can be defined as "the amount of waste that one person throws out during a certain period in terms of weight or volume." Weight is often used because its change is much less than the change in volume during transportation. The reason is that the weight is not greatly affected by the degree of pressure to which the waste is exposed [1]. Many factors affect the quantity and quality of solid waste generated in cities, the most important of which is [1]. First, the country's economic direction, whether agricultural or industrial, In agricultural countries, the majority of waste generated is organic matter that is degradable and rots because it contains a high percentage of food waste and agricultural waste. In industrial countries, the largest proportion of waste comes from manufactured materials such as glass, metal cans, and others. Second, the economic and livelihood level of the country affects the increase in the amount of solid waste. Third, population density, as the waste increases with the increase in the population. Fourth, the amount of generated waste is affected by the type of service provided by the municipality or the service sector in the field of solid waste management, including methods of collection, transportation, and treatment.

6. Waste flow stages:

The solid waste flow goes through six stages, which are [10]:

The first stage is prevention, which is related to the waste policy rather than the actual waste treatment.

The second stage is the generation or formation of the waste in which the origin of the specific waste is the owners of homes, industry, hospitals, commercial workers, and public units. The third stage is reuse or recovery. However, there are several reasons for recovery, including reducing the amount of waste sent for final disposal, thus reducing the need for transportation and disposal; increasing revenues through the sale of recycled materials; and reducing the use of virgin raw materials. The fourth stage is the waste collection, which is applied only to some of the waste generated by producers and mainly applies to waste from homes and small businesses, and this stage also includes returning the products to the source.

The fifth stage is transporting and exporting the waste from the originator to a proper site for treatment. But, given the special characteristics of hazardous waste, special precautions must be taken during collection and transportation, including the training of the driver and his assistant, the types of packing used, the marking of the packing, and the transport vehicle.

The sixth stage is waste treatment and disposal, often in the same natural place. Disposing of the waste in pits is the normal solution to treat each of the waste or the residual waste that cannot be treated as part of other means of waste treatment, such as converting them to compost, ash, or casting.

7. Solid Waste Management Concept:

Waste management can be defined as a term that describes several distinct processes, which include the process of solid waste collection, disposal or transportation, recycling, and final treatment of waste by removing toxins, making it permanently harmless to water, air, and land, which is one of the biggest challenges facing developing countries [4, 10].

The increasing population density, especially in urban areas, as well as the diversity of waste sources add to the difficulty of adapting to the accumulation of this waste in residential areas, which leads to the necessity of establishing waste disposal systems regularly. There are specialized waste departments, especially in industrially developed countries. During this period, there were many variables, and the high standard of living was an important factor in the generation of waste [9].

Solid waste management in any city is one of the indispensable foundations for organizing, controlling, and dealing with solid waste, as this waste poses many health risks in its various stages of existence. The importance of ensuring good solid waste management is now recognized at all levels, whether at the international, national, or community level, as an essential component of sustainable development. Therefore, solid waste management is a mandatory function of urban local authorities [2].

To design an appropriate waste management system, the following objectives must be achieved [1]. The first is the protection of public health. Second, achieving a high-quality urban environment; third, supporting the economy; and fourth, providing job opportunities.

As for the concept of integrated solid waste management, it is an approach or method responsible for the failures of traditional solid waste management. It was established and initiated in the United States of America by the Environmental Protection Agency (EPA) in the early 1990s to expand the solid waste management system instead of focusing on waste disposal. It also includes preventing and reducing waste to the minimum source, reusing and recycling it. It is also a social, environmental, and institutional culture that effectively affects the continuity (sustainability) of waste management and depends on long-term strategies [1].

However, the main objective of integrated management is to reduce the amount of waste that is disposed of in sanitary landfill sites, as the principle of the four-way rule achieves the following objectives [1]. First, enhance cooperation between waste producers and waste collectors who work in the processing and manufacturing stages. Second, reduce environmental degradation. Third, save energy, reduce the depletion of natural resources, and preserve them to achieve sustainability. Fourth, good solid waste management.

The integrated management of solid waste also includes a set of plans and strategies that help achieve its goals and objectives, which [1] are: first, minimizing the source; second, recycling; third, waste transportation; and fourth, landfill. These strategies are hierarchically shaped as in Figure (1).

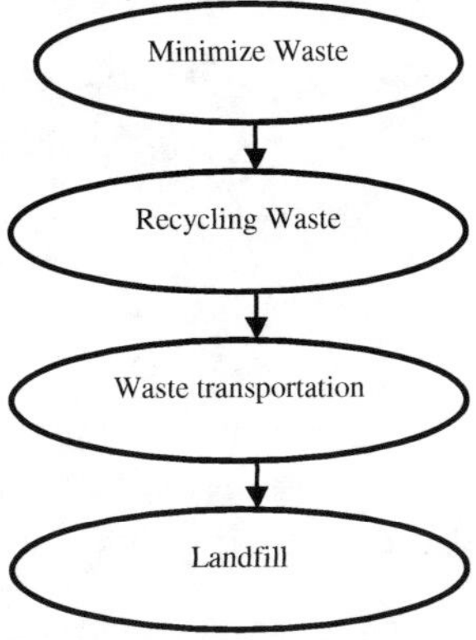

Figure 1. Integrated Solid Waste Management Strategies

Integrated solid waste management focuses on the principle of using the four golden rules, which include four strategies (reduction, reuse, recycling, and recovery) according to a hierarchy of management strategies to reduce waste, as shown in figure (2)

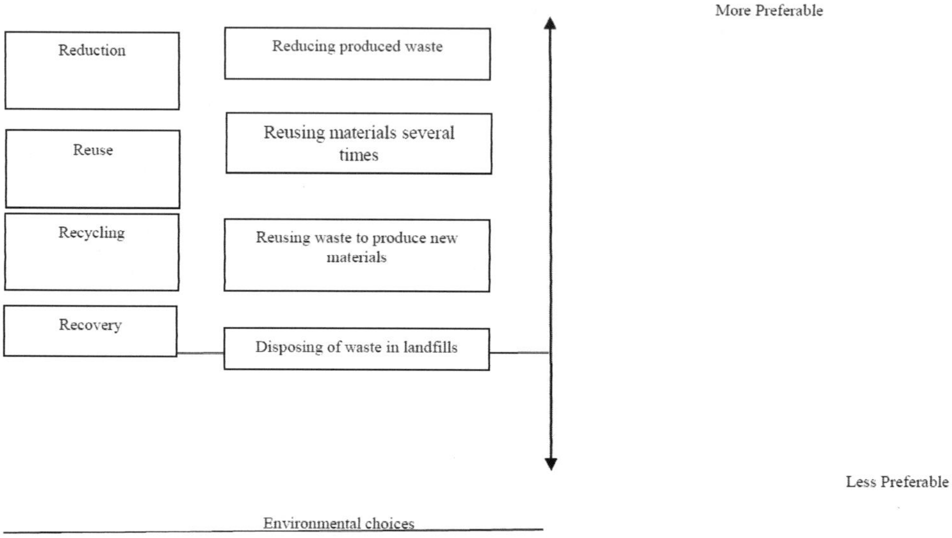

[10].

Figure 2: the hierarchy of solid waste management

8. Modern waste management:

The solid waste management hierarchy sets waste management options according to environmental preferences. The important priority is to prevent the generation of this waste, and this is considered first. When waste is generated, the priority will be for the process of reuse, then recycling, then recovery, and finally disposal of it in a landfill. Waste prevention: encourage the use of products that are durable, safer, and can be reused for longer periods. Reuse encourages the reuse of materials without the need for further processing, as in partial or complete cleaning and repair of materials. Recycling concerns treating waste and then converting it into a new product. Waste recycling protects the environment and natural resources. Other activities for recovery include reuse, recycling, and treatment, as well as the optimum use of the resources recovered from waste. From the observation of the hierarchy, it is clear that waste disposal is the most appropriate option, especially for some types of waste that cannot be recycled, as shown in figure (3) [1].

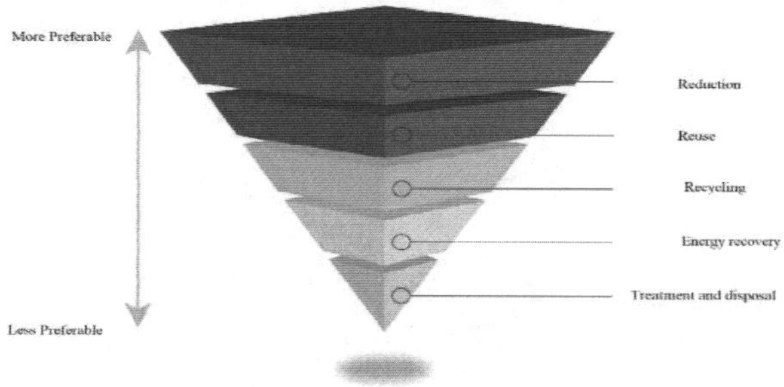

Figure 3: Waste Management Hierarchy

9. Safe Environmental Management Strategies for Solid Waste:

Among the most important of these strategies, which are the best solutions to solve the solid waste problem, are as follows [3]:

1. Using simple methods to reduce waste production. This method is dependent on the citizen's and society's level of awareness.
2. reusing waste, such as reusing one of the materials for the same purpose several times or reusing the materials by using them for new purposes.
3. Incineration of waste using modern incinerators capable of controlling air pollution.
4. The use of sanitary landfills as a necessary method of burying non-combustible or recyclable waste.
5. Handling and burying hazardous solid waste in designated areas
6. Organizing awareness and information campaigns for a variety of societal sectors
7. Research, development, and education.

There are other strategies, the most important of which are [1]:

1. Strategy One, Zero Waste: Supporting Zero Waste Ethics through Metro Fanacker's Social Marketing, Communication, and Education Programs The amount of waste you produce is directly related to the number of goods and services that are consumed.
2. The second strategy is called "consumer and producer sharing responsibility." The strategy of consumer and producer sharing responsibility and arguing that governments should transfer additional waste management responsibilities to the producer and consumer is not new. That is, the costs and responsibilities of waste management are borne by local governments and the taxpayer, and the costs and risks of managing end-of-goods must be the responsibility of the producer and consumer who uses them, not the taxpayer.
3. The third strategy is the waste reduction strategy. The center adopts this strategy to reduce waste, such as reducing waste wood, increasing the opportunity for wood waste recycling, and providing wood waste collection organizations for recycling.
4. The fourth strategy is to reduce the waste of paper and paperboard. About 14% of the waste disposal in America consists of paper and adds reference, and most of it should enter current recycling programs. Contaminated paper that can be recycled can be decomposed with other organic materials to produce reusable products.
5. The fifth strategy, targeting organic materials for recovery, Food waste represents 13% of the waste presented. These materials can be degraded together with garden and patio waste and some cardboard to produce useful and marketable products.
6. The sixth strategy, the waste plastics recycling strategy, entails expanding the process of collecting recyclable plastics by the residential consumer and encouraging more plastic recycling operations in the commercial sector.
7. The seventh strategy: This strategy represents a focus on densely populated areas and targets multiple sectors of housing or families to improve the rates of legislation.
8. The eighth strategy is the strategy of establishing the infrastructure for the conversion of waste into energy through expanding its infrastructure. The waste streams are treated on-site in a sustainable manner that provides benefits in the form of energy and heat to the region.
9. Strategy Nine, Waste Metal Recycling System Development Strategy: Develop a foundry (metalworking) ash recycling system showing the possibility of recycling ash from the Metro Vancouver Center in the USA by grinding, sequestering, and removing metals and other ore residues into compatible regular products (piece one).
10. The tenth strategy is the disposal of residual waste and minimizing the environmental impact. Seeking to search for the best options for the possibility of disposal and the optimal use of disposal of treated waste that is no longer useful.

10. Stages of solid waste management

1. The stage of generating flexible waste: The generation of waste is inevitable as a result of repeated publishing activities, and it comes in different quantities depending on the conditions of individuals, their standard of living, and the degree of culture that characterizes the society. Attention should be paid to the quantities of waste generated, as they cause waste and excessive use of non-industrial resources, which will lead to disruption of their existence. Therefore, the issue of waste generation is one of the priorities that studies and research focus on. They focus on solid waste treatment to reduce waste generation and use several alternatives, the most important of which is the principle of recovering some resources from waste. The habits and traditions of individuals must be changed voluntarily and voluntarily in a direction that reduces the generation of solid waste, and this is done by spreading economic and environmental awareness by different means. The phases of solid waste management include homes, businesses, service organizations, and shopping malls. The government can control or influence the amount of waste through material incentives. By the principle (who polluter pays," the incentive is the obligation imposed on producers to substitute less harmful hazardous materials [4, 7].

It is the first problem with the resulting waste, as all subsequent stages depend on it. If the processes of sorting and isolating them from the source are carried out, the successive activities that follow will be facilitated. Estimating the appropriate size of waste storage containers depends on the rate of their generation for each individual, the number of family members and the type and nature of the waste presented. The most important types of storage containers are plastic and iron containers [6].

2. Solid waste storage stage: The main purpose of solid waste storage is the safety of the general environment from disease vectors in addition to the aesthetic aspect of the city. Estimating the appropriate size of waste storage containers depends on the rate of solid waste generation for each individual, the number of family members, and the type and nature of the waste. Steel containers with capacities ranging from 5–70 liters, bags made of real plastic and used only once, and collective storage methods are the most important types of storage containers [4].

The flexible waste storage phase follows multiple methods as it is collected manually and through the fixed and mobile containers. The importance of this stage lies in preserving nature, preserving public safety and aesthetic value, increasing productivity, ease of handling containers, and reducing collection costs. It also has an important role in organizing and removing solid waste. There are important differences in solid waste storage methods according to income levels and social conditions. For example, in high-income population sites, plastic or metal containers are used according to international characteristics, but in low-income areas, waste is stored in the form of piles, baskets, or cages [4].

3. Solid waste collection stage: The process of solid waste collection, especially in urban areas, is a complex and difficult process due to the various activities and events and the variation in the areas in which waste is generated and the collection process. The process is defined as picking up waste and placing it in vehicles designated for collection operations. The process starts from the time the vehicle arrives at the waste generation areas until it completes its daily route. Collection operations need workers, supervisors, mechanisms, and requirements through transitional stages, according to those means and methods that are affected by several and manifold data, including the social and economic reality of the city or area, the specific area in which the collection takes place, as well as the urban design and land uses for those cities. For example, waste collection methods in vertical buildings differ from those in open horizontal areas as well as in cities whose alleys are narrow. At this stage, the storage containers are transferred to large collection stations or directly to the final disposal sites. The collection process is carried out using different collection techniques [1]:

1. Containers: Containers used to store waste in areas where waste is generated must be strong, easy to transport, and economical, as well as anti-corrosion. The containers that are used for collection operations in residential areas are metal or plastic and are designed according to the automated system in force in the transportation mechanism.

2. Frequency of collection: The conditions of the climate and the area in which waste is generated determine the number of collection times. In a hot and humid climate, the number of collection times is twice a week due to the decomposition of solid waste and the emission of foul odors from it, especially residential waste, most of which is food waste. As well as the type of container used for collection, for example, closed containers help to collect three times a week, and open containers collect daily. Finally, the efficiency of the collection depends mainly on the demographics of the region.

The waste collection process includes several stages, as follows [4]:

a.We start at the house level, which is called the initial collection, and it is a very important stage as it is the first link in the collection chain. In which the waste generator collects his waste and puts it in the collection receptacles in his residence or puts it in a bag to be presented outside for disposal.

b. The municipality collects waste collected in public places.

What is spent on solid waste collection activities is approximately (50-70)% of the total amount of money spent on organizing inflexible waste, because a large part of the total cost is linked to the collection process. Thus, any improvement, even a small percentage, in the collection process can lead to significant savings in the cost of the system and in general.

The stage of collecting the products generated from their various sources is of importance in the sequence of steps so that the rest of the subsequent stages depend on it [6].

a. means of transportation: It is the fourth stage in waste management operations and includes the transportation of solid waste generated in various places through two stages:

First, transport it from the house to the large containers that are usually scattered in predetermined places on the streets so that it is easy to transport.

Second, is the stage of transportation to the treatment and final disposal areas, which is carried out by large containers or specialized mechanisms.

b. Transfer stations: Specialized mechanisms are used in places designated for waste collection. Since the quantities of household waste are increasing and the distances they are transported to the landfill or treatment sites are large, the cost of transporting them there will be high if the same collection mechanisms are used for transportation. But the costs will be lower if transformational stations are used, as in these stations the waste will be collected and placed in large-sized machinery or containers and transported to the final sites or transported by rail. Also, these stations have waste compressors, which help reduce transportation costs, as larger quantities will be transported at the same fee.
The concept of (transformational stations) is a compromise solution to the process of collecting and transporting solid waste together to rid certain areas within the city of waste accumulations that, if they remain for a long time, cause odors, insects, and everything else that harms the environment of those areas. Transformation stations are usually resorted to when using small machinery that collects waste from narrow areas and alleys, especially in traditional cities, and sends it to these stations to transport it later with larger mechanisms to the final sites or treatment sites, whether they are sanitary landfills or reprocessing plants. Transfer stations play an important role in solid waste management in many societies, as they are the link between the collection system and the final disposal sites for waste or treatment plants. The main reason for establishing intermediate stations is to reduce the cost of transportation to treatment plants or sanitary landfills in terms of time, fuel consumption, and maintenance of machinery and equipment [1,7].
4. The stage of solid waste transfer: the main objective of this process is to transfer waste from the sources that generate it, such as residential or commercial, and all waste-producing activities (various temporary storage points) and deliver it to special places such as intermediate stations or safe burial sites for burial or treatment. It is the responsibility of the municipal council to ensure the cleanliness of the city and to get rid of its environmental impacts, and this is done through the transport fleet maintained by the competent local authority. It is very necessary to synchronize the entire waste collection process with the means of solid waste transportation. It includes sorting or recycling operations and converting them into energy, according to the country's level of progress. Waste is collected from storage and collection containers for solid waste treatment and permanently removed, and this waste is usually transported by different mechanisms [4]. The transfer phase involves two steps, as follows [4]:
a.Transfer waste from a small waste collection vehicle to large transportation equipment.
b. Subsequent long-distance transport of waste to a treatment facility or disposal site.
5. Solid waste treatment stage: The need for adequate treatment and waste disposal arose when the population began to move from dispersed residential areas to collective residential areas, forming local communities. The increase in the number of people in countries and cities resulted in an increase in the amount of waste generated by them, which became a troublesome problem. Waste was burned as a disposal method at that time. The goal of treating waste before it reaches the landfill is to prevent or reduce the harmful effects of the landfill site on the environment. Several directives were issued regarding dealing with waste, including the (European Union) directive in 2008, which formed a framework on how to deal with and manage waste, and the goal was to reduce the amount of waste at landfills. Waste is buried in a scientific manner in which the environmental aspects of the place are preserved. The old methods of waste collection in the cities of Iraq in general and the study area, in particular, depend on garbage bins and yards that work on paved roads and open collection points in areas that are difficult to penetrate, where individuals take it upon themselves to carry their waste to areas where it is easy for trucks to collect waste. Many methods are used to treat large quantities of waste, and each of these methods has its advantages and disadvantages, whether about cost, means, required space, or impact on the environment [4].

11. Solid Waste Management Audit:

It is a formal structural process used to measure the amount and types of waste generated by an organization. The information used from the audit helps identify current waste practices and how they can be improved. This means [10] the following:
1. Organizational performance efficiency and effectiveness
2. Lowering the cost of solid waste disposal.
3. Economic in the use of limited natural resources.

12. Objectives of Solid Waste Audit:

The objectives of a solid waste audit can be defined as follows [10]:
1. Determining the composition, characteristics, and quantity of waste generated by the activity.
2. Assessing the efficacy of existing waste management systems.

3. Look for ways to improve waste management systems and strategies.
4. Collect basic data to measure the effectiveness of waste reduction strategies.
5. Examine the waste tracks.
6. Gathering detailed information on waste generation and production costs.

13. Waste collection methods

It means solid waste collection operations from areas with different uses and activities. These processes require work, supervisors, mechanisms, and requirements through transitional stages according to the methods and methods that are affected by many and complex data, including the social reality of the city, area, or specific area in which the collection is carried out, as well as the urban design of those areas. For example, the methods of collecting waste from vertical buildings differ from those of horizontal open spaces as well as areas located in traditional cities with narrow alleys [1]. They are in several ways, as follows:

The method of joint addition includes the following [1]:

a. Waste disposal in designated areas: Waste is collected in designated, designed, and closed waste collection areas. This method is less expensive, but waste dumping is slow and unhealthy due to the spread of waste around the collection area and the emission of unpleasant odors.

b. Shared containers: Waste is collected inside large and shared containers that are unloaded from waste collection vehicles and also have a low cost. One of the disadvantages of this method is that the containers need to be maintained on an ongoing basis or replaced due to damage or oxidation.

2. The method of singular addition is as follows [1]:

a.Collection from the residential block: the collection is carried out through the movement of waste transport vehicles in shops and areas to collect containers after placing them outside the homes. One of its disadvantages is that the containers are left for long periods until the arrival of the collection vehicles, and this leads to the spread of waste in the streets because of the animals.

b. Collection from the sidewalk: Waste is placed outside the building in containers or placed in plastic bags for collection or unloading. Among its benefits, all waste is collected, and its harms are the spread of waste in the streets due to wind, animals, and children.

c.Collection from homes: Waste is collected by workers in the field of waste by informing the residents to collect waste, and this method is suitable for the residents but is not useful for vertical housing.

14. Collection systems for solid waste

It includes the following [1,7]:

1. Portable container system: In this type of collection system, the containers used for storing waste are withdrawn to the treatment area, transfer station, or landfill area, where their cargo is unloaded and returned to their original location.
2. Portable container system-site switching system: the empty container is moved from the sites of the transfer station or the sanitary landfill site to another site. In this type, it is easier to work with when the containers are small, as the beginning of the path is from the container unloading site to other waste collection sites.
3. Fixed container system: The location of the container is fixed in the waste generation area as the path of the transport vehicles is towards the fixed container sites to be unloaded.

15. Solid waste sorting and recycling stage

Waste sorting and recycling represents a group of operations that include collection and treatment of waste and the production of raw materials that are used in the manufacture of the same product from which the waste was generated, or another product such as fogged glass that is used in the manufacture of various other glass materials. What is meant by recyclability is the extent to which it is possible to benefit from waste that is supposed to be disposed of by any known means of disposal. And the waste, from the point of view of its producer, is of no value, and from the point of view of the environment, any action is taken to re-utilize this waste, whatever the cost, is a beneficial process. It is a great benefit, at least a decrease in the amount of waste injected into the environment, and this in itself is again. Therefore, recyclability means the extent to which a raw material can be recovered from a waste that can be used as a raw material in the production of the materials from which the same waste material was produced [1]. Accordingly, the following things must be taken [1]:

1. The waste should be easily accessible and easy to separate.
2. The specifications of the raw materials in the waste are recoverable.

Among the strategic dimensions of the recycling process are the following [1]:

1. Energy saving: companies investing in energy, as it saves time, cost, and effort in saving raw materials. For example, as the material resulting from the recycling process is an alternative to the raw material, two-thirds of the iron industry in the United States of America comes from recycling used iron, and this saves 74% of the energy consumed. Similarly, 51% of the paper industry is made from recycled paper, as is 80% of the plastic industry.
2. Preserving natural resources: Many industries depend on natural resources, and the demand for these resources continuously will reach depletion within a period. For example, recycling a ton of waste paper can protect 17 trees.
3. Environmental Protection: Efforts are being made to reduce the environmental impact of uncirculated solid waste.
4. The economic dimension: It constitutes a low-cost investment opportunity that achieves rewarding financial returns and does not require large funds, providing job opportunities and providing raw materials for many factories at a lower price than if they were new or natural raw materials.

16. Solid waste disposal stage

There are two ways to dispose of solid waste, which are as follows [1]:
1. Landfilling (landfill): Landfill is a method of getting rid of solid waste by using specific land without causing damage and dangerous effects on the environment and public health. For landfill land, clay or artificial layers and barriers must be used to control the discharge of water resulting from the waste leachate. After the land is cleared of the waste from the burial of waste, it is planted with grass and hashish to turn it into a garden. as well as a control system for methane and leachate generated from the landfill. The land is selected according to appropriate engineering methods by health and environmental specifications. The waste is compressed to cover the least possible area. Landfill sites are usually outside the boundaries of urban areas, which include water treatment plants, agricultural areas, main roads, etc. Because these sites are incubators for many insects, such as mosquitoes and rodents, they cause the transmission of many diseases to humans [1,7]. Among the advantages of burial are the following [1]:
Preserving the environment and public health.
Avoiding the spread of rodents, insects, diseases, and fires.
maintaining the overall appearance.
An outlet can be made for methane and other gases to be used as fuel.
The purpose of the landfill site design is [1]:
1. reducing the amount of leachate produced at these sites by constructing an insulating layer that keeps it from entering groundwater and soil layers.
reducing the quantity of hazardous waste.
The burial produced the following results [1]:
1. Leachate is a liquid that collects from water sediments and contains chemicals resulting from the vital reactions of landfills.
The gases emitted from the landfill are a mixture of methane (CH_4) and carbon dioxide (CO_2) produced by the decomposition of organic matter in the solid waste inside the landfill. The rest of the gases generated from the landfill are nitrogen, oxygen, and ammonia.
Waste incineration: is a way to reduce the volume and weight of solid waste with or without a heat treatment system using devices that help treat the ash generated from the combustion process and environmental control devices. The incineration of solid waste takes place inside modern furnaces equipped with modern devices to control the gases emitted from them, such as toxic carbon dioxide. The purpose of solid waste incinerators is to reduce the volume of waste by 90%, and the rest of the parts, such as glass, metals, and ash resulting from the combustion process, are dealt with in other stages. Fly ash and combustion gases are treated as hazardous waste. The burning of waste under controlled conditions is usually carried out in closed areas or buildings, and it can be defined as the pneumatic and thermal treatment of solid waste, whether by generating or not generating energy and getting rid of the incineration residues. It is carried out by a series of operations within specialized laboratories [1].

17. Methods of dealing with inflexible waste

There is a set of methods that can be used to deal with solid waste generated from various sources. The most prominent of these methods are:
Preventing or reducing waste generation at source: Waste reduction is a term synonymous with waste prevention or waste minimization. It has been defined as a method, process, or activity that eliminates or reduces waste from

its source. Waste reduction is both environmentally and economically beneficial to society as a whole. The best way to implement this method is through awareness campaigns specializing in this field. It is also possible to reduce the rate of waste growth and make people change their environmental behavior through the application of economic tools, such as assigning citizens to pay part of the cost of raising waste in the form of fees at certain percentages with the issuance of appropriate legislation. This is approved in many countries of the world and will reduce people's consumption of goods and the amount of waste generated by them [4].
Reuse: Human activities generate solid waste, which is often disposed of because it is considered useless. The term "waste" suggests that materials are useless and unpleasant, yet many of these wastes can be used. Thus, it can become a source of production or energy generation if it is managed properly, and reuse is more effective when there is a separation from the source, where the separation at the source leads to the following [4]:
Increasing the percentage of waste recycled by preventing the mixing of recyclable materials with organic waste and improving the quality of the collected non-recyclable materials.
By reducing the total amount of waste, we can extend the average life expectancy of the burial site.
They play an important role in raising public awareness.
d.Increasing the economy's value
lower freight costs and shipping-related issues such as noise, air pollution, and traffic congestion.
3. Recycling: It refers to the treatment of waste materials to reuse for their original purpose or other purposes, including the recycling of organic materials. Recycling is the process of separating and collecting secondary materials for recycling and is considered an important economic factor, providing income opportunities for the poor, the unemployed, and those with special needs. It significantly reduces the amount of waste that has to be collected, transported, and disposed of [4].
4. Recovery (e.g. energy recovery): the conversion of waste into energy through the use of controlled facilities to incinerate solid waste that significantly reduces its volume. Municipal facilities that convert waste into energy have many advantages over waste management systems. Combustion can destroy bacteria and viruses in the waste as well as harmful organic compounds. It also reduces the volume of solid waste by up to 90%, thus conserving the space of the waste burial place. This type of treatment has been used for a long time to reduce solid waste, reduce transportation costs to treatment sites, and absorb waste for more people [4].
5. Disposal (sanitary landfill): Landfill is one of the simplest and cheapest ways to dispose of waste, and for this reason, it was the most common method in the world when suitable sites were available at a reasonable distance [5]. Placing solid waste on the ground is termed "discharging." And open discharge is the least costly means of solid waste disposal. The method was used by the majority of societies in the past. Often, the volume of the tailings at the discharge site is reduced by incineration, thus extending the life of the discharge. But the unloading site suffers from a large number of rodents, the appearance of odors, air pollution, and the presence of insects, and can cause serious environmental damage. As a result, the search for alternative methods led to the development of unloading to a safe landfill site. As shown in figure (4) [4], sanitary landfills differ markedly from open landfills, as the latter are just places to dump waste, while sanitary landfills are engineering processes that have been designed and operated according to acceptable standards, as shown in figure (4) [4].

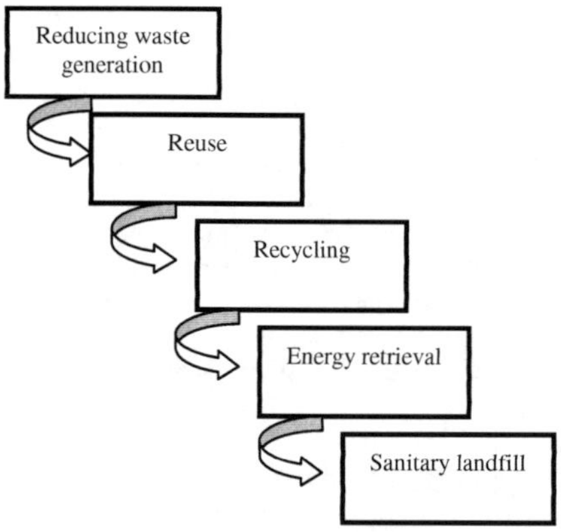

Figure 4 . Methods for dealing with waste

18. Human characteristics affecting solid waste service

The study of the characteristics of the human study area is an important part when studying any phenomenon in the geography of cities, which are as follows [11]:

Population size: overpopulation is more dangerous than pollution in itself because it is its source. As it contributes to the deterioration of services and service facilities and has an impact on natural resources identified through solid waste from them, Knowing the number of the population is a basis for knowing the extent of its compatibility with the service provided and the nature of the continuous demand for the service, with an understanding of the behavior of the solid waste service phenomenon, so that the planning process is based on scientific foundations to achieve efficiency and adequacy of the service provided. The population increase is also one of the most important factors affecting the increase in solid waste, as it is directly proportional to the increase in the quality and quantity of solid waste presented by the population. This is related to the amount of solid waste that the individual raises daily, which varies according to the economic level, their number, and their cultural level.

Because of the economic reality and the circumstances that Iraq has gone through, some housing units are inhabited by more than one family. The number of households was adopted as one of the factors affecting the solid waste collection service.

Population density is crude: it is the simplest type of measure used in population studies and simply means the total number of people in an area unit.

The area of the municipal unit: the area of the municipal unit is one of the factors that affect services, and the relationship between the city's area and the need for infrastructure services is complex and determined by several factors. The larger the area of the city, the greater the potential for savings in many types of services, which leads to a reduction in the cost of the unit provided. Two main factors stand out in the area of the municipal unit and its relationship to waste collection: The first factor is in calculating the time taken to collect waste, which affects the number of mechanisms and the cost of collection.

The standard of living of the family: The standard of living of the family is one of the important human factors affecting the quantity and quality of solid waste. The increase in economic growth rates and the rise in the average income of the individual and the family lead to an increase in the demand for infrastructure services, as the ability of the family to own goods that they were unable to own at a low level of income increases. The higher the standard of living, the higher the rate of waste generation. The level of income affects two aspects: the first aspect is direct, as with an increase in income comes an increase in the volume of waste that is thrown up, while the other aspect is the decrease in income, which means the family's ability to acquire containers or waste bags, which causes them to be thrown in the open.

The educational level of the family: The study of the educational structure is of great importance in demographic and urban studies within the city, where the educational structure data reveals the degree of development and scientific progress experienced by the society.

There are other factors, including [12]:

The distance between the container and the house: The distance of the container from the house is one of the factors affecting the accumulation of waste and hurts the process of collecting it. In general, the greater the distance between the house and the container, the less the citizen throws waste into the container, and the closer the container is to the house, the more the citizen is inclined to throw his waste there.

Solid waste sorting: The waste management process is affected by citizen sorting of solid waste into many components, as the majority of the population deals with waste as a single mass. This process has great risks, as this stage is one of the most important in reducing the volume of solid household waste. This step affects the mixing of organic household waste with each other and provides it with an opportunity for fermentation, which results in many problems, such as unpleasant odors and the spread of rodents, which are a cause of disease transmission to the population and directly to the workers in collecting and transporting waste.

19.Natural factors affect municipal solid waste service.

The following is a review of the most important natural factors affecting municipal solid waste service:

Location and location: The city's geographical location is of great importance when studying the municipal solid waste service, because, by knowing its characteristics, it can know the most appropriate ways to manage these services with the least effort and cost. The location differs from the location, as it is an area, not a point, and it is the justification for the existence and continuity of the city. The site is specific to the city, while the village and the city share the same geographical position. It is one of the most important elements of the natural environment that has a pioneering effect in shaping the characteristics of any region and giving it character. In addition to

being a key factor in showing the effects of the astronomical location on the longitude and latitude of any region or region, these influences have a role in highlighting the climatic conditions of any region in the world [13].
Geological composition and soil: It is known that organic and inorganic materials are present in solid waste. As well as the materials resulting from the decomposition process can seep (leached) into the soil using rainwater or surface water that falls or moves through the places where the waste is outside the containers, which means its pollution, thus polluting the groundwater. The effect of leaching depends on the quality of the waste (hazardous or non-hazardous), its quantity, and the amount of water it receives, in addition to the texture of the soil. A coarse texture does not retain much water compared to fine-textured soil. Also, these resulting materials may descend into the already polluted waterways and increase their pollution [13].
Climate: The climate, with its various elements, has a significant impact on municipal solid household waste and its management as it affects the processes of collection, transportation, and landfilling because it either contributes to an increase or decrease in the population's waste output when temperatures drop and rise; or that some of them are affected by the seasons of the year; or that they scatter or move quantities of them and sweep them from their places to other places when rain falls or winds blow; as well as contributes to the decomposition of some of those waste components. Some of them may also cause fires in their collection areas or sanitary landfill sites before burying and permanently disposing of them, and they may even have some control over the working hours of the cleaners themselves [13].

20.Environmental damage from solid waste

Among the most important are the following:
The requirements of public health care require the prompt removal of all waste from populated areas, and its disposal without harmful effects [1].
Waste and solid waste cause great damage to the environment, and open garbage may be responsible for many diseases. Incinerators add pollution to air pollution, and random methods of waste disposal contribute to harming the environment. There are many risks caused by solid waste, as it contributes to As a result, waste is attracting huge numbers of insects, especially cockroaches, which transmit serious diseases to humans, as waste provides the appropriate heat and humidity in addition to the food suitable for breeding many generations of these insects, and waste contributes to visual pollution, air pollution as a result of burning waste, and pollution of groundwater and drinking water as a result of dumping it into water sources. But the recognition of the impact of waste on the environment came relatively late, as many cases of water poisoning by leachate from landfills are generally recognized, along with the risk of explosions and the toxic effects of air particles on those living in the middle of landfills [6, 10].
Aesthetic and social effects: it causes harassment and psychological effects as a result of visual pollution, aesthetic damage to the city, and the citizens' psychological health [1, 10].
As the accumulation and accumulation of solid waste in an area causes its distortion from an aesthetic point of view, causes distress and disgust in the mind of the beholder, and thus affects the residents of the region psychologically, the negative effects of solid waste generated are not limited to health, environmental, and economic aspects, but the matter includes the same aspects. and the decline in services provided by the municipality, which led to huge accumulations of waste, including in residential communities and public stores [6].

The absence of vision and future planning for the expansion of residential neighborhoods or agricultural areas and the use of deserts as open dumps for solid waste, especially in areas that have fertile agricultural land, have wide pastures for animals or represent drinking water from wells, in addition to those areas with desert nature and tourism, or on tourism routes and important monuments [6].
Waste contains minerals, potential energy, and other useful materials, and the process of recovering them is an economic benefit. Wasting these materials also means an economic loss for the country [1]. The European Environment Agency explained that the generation of waste reflects the loss of materials and energy and causes economic and environmental costs to society through the process of collection, treatment, and disposal of waste [6].

21-Conclusions:

The research reached several important results, as follows:
1-Solid waste is any material that is thrown away by a person because it is no longer needed and is not usable.
2-Among the most important types of solid waste, as well as street and other waste.

3-Municipal solid waste is one of the materials that can be recycled, the most important of which are organic, metal, paper, cloth, and plastic.
4-One of the most important factors that affect the quantity and quality of waste are the country's economic trend, whether it is agricultural or industrial, the country's economic and livelihood level, and population density. The amount of waste generated is affected by the type of service provided by the municipality.
5-Waste compositions consist of physical and chemical compositions.
6-Among the most important objectives of solid waste auditing are: determining the composition, characteristics, and quantity of waste generated by the activity, measuring the effectiveness of the existing waste management systems, and others.
7-The concept of waste management is a term that describes several distinct processes and includes the process of collecting and disposing of solid waste or transporting and recycling it.
8-In order to design an appropriate waste management system, the following objectives must be achieved: protecting public health, achieving a high-quality fertile environment, supporting the economy, and providing job opportunities.
9-Integrated solid waste management includes a set of plans that help achieve its objectives: source reduction and recycling, waste transportation, and waste obliteration.
However, among the most important stages of waste management are the stage of generating flexible waste, the stage of solid waste storage, the stage of solid waste collection, and the stage of solid waste transportation.

References

[1] Hammoud S, Reda H (2016) Developing a solid waste management system (applied research in the city of Hilla). *Journal of Economic and Administrative Sciences*, **22** (91), pp. 31-138.
[2] Abd J, Muhammad R (2018) A geographical analysis of the reality of solid waste management in the city of Hilla. *Journal of Human Sciences*, **25**(2), pp. 86-92.
[3] Abdullahv S, Ali, A (2011) The extent of applying safe environmental management strategies for solid waste: an exploratory study of the opinions of a sample of administrative leaders in the Mosul municipality directorate. *Rafidain Development*, **33**(105), pp. 192-198.
[4] Jamil R, Jarallah F (2020) The role of solid waste management in addressing the environmental pollution crisis, a field study in the Mosul Municipality Directorate. *Tikrit Journal of Administrative and Economic Sciences*, Fourth Scientific Conference: The Hidden Economy and Crisis Management, **61**(Special Issue), **61**, Part 6, pp.481-488.
[5] Thamer A, Ahmed M (2009) A new method of solid waste treatment in the city of Mosul and its cost comparison with the current method of waste treatment. Tikrit Journal of Engineering Sciences, 16(4),42.
[6] Hamed F, Muhammad F (2020) Environmental pollution by solid waste and the methods used in its management, a case study in the municipality of Ramadi. Journal of Economics and Administrative Sciences,26(122), pp.148-152.
[7] Abdul Latif N, Abdul Muttalib, H (2005) Evaluation and development of municipal solid waste management in the city of Najaf. Engineering Journal, 11(4), pp.71-76.
[8] Ali M, Farouk G (2007)Solid waste management and treatment in some Baghdad city hospitals. Engineering and Technology Journal, 25 (5), pp. 232.
[9] Hamad W, Mohammed S (2020) Municipal solid waste in Iraq between the hammer of population growth and the anvil of management: facts, treatments and a vision for the future. The Iraqi Journal of Market Research and Consumer Protection, 12 (1), pp. 74-75, 82.
[10] Abdel Rahman V, Saad M (2018) The impact of environmental auditing on solid waste management in maintaining a sound environment and reducing negative effects on the environment (an applied study in the Al-Ghadir Municipality Department - Baghdad). Journal of Baghdad College of University Economics, (54), pp. 375-377.
[11] Hashem A (2021) Spatial variation of solid waste service in the city of Baghdad (an applied study in the municipality of Al-Sadr first and session). Journal of Human Sciences, 28(1), pp.5-10.
[12] Hussein N (2019) Abdullah S, Evaluation of Solid Waste Management in Tikrit City. Adab Al-Farahidi Journal, 11(37),pp. 345.
[13] Lafta M, Ismael O (2021) Natural factors affecting the management of municipal solid waste service in the city of Basra in 2018, Basra Research Journal for Human Sciences, 54(4), pp. 186-189.

Air Pollution Resulted from Coal, Oil and Gas Firing in Thermal Power Plants and Treatment: A Review

Rafid M. Hannun[1] and Ali H. Abdul Razzaq[2]

[1,2]Department of Mechanical Engineering, College of Engineering, Dhi Qar University, Dhi Qar, Iraq

[2]Email: alhba7746@gmail.com

Abstract

Air pollution is a worldwide and a local issue caused by energy generation. It refers to global warming deterioration in human environmental health and local-global sustainability. This review provides information on the generation and consumption of energy with their air pollution mixture of many pollutants; gases, liquids and particles. These pollutants have become one of the leading environmental hazards to human and planet health. The components of pollutants that result from burning fossil fuels (oil and gas) and coal have been studied, such as sulfur dioxide, carbon dioxide, nitrogen oxides and particles. Both sulfur dioxide and nitrogen oxides interact with water to produce acid rain. Both gas and oil-fired thermal power plants emission today and in the future might be transformed to work on renewable energy sources. The results stated that gas-fired power plants are more generally faster efficient, less pollution than oil and coal power plants. A complete assessment of exhaust gas treatment was conducted. Carbon dioxide capture, desulfurization, denitrification, and particle collection were all successful as treatment control mechanisms. So, previous studies presented some methods for reducing nitrogen oxides and sulfur oxides through water injection and exhaust gas recycling. The addition of nano-additives in diesel emulsion fuel (W/D) has recently been shown to increase the characteristics and performance of the fuel while reducing the quantity of emitted hydrogen chloride (HC) and carbon monoxide (CO). It was determined that the best nano-additives for W/D were Aluminum Oxide (Al2O3), copper(ii) oxide, magnesium oxide (MgO), manganese oxide (MnO), and zinc oxide (ZnO), among others (E10).

Keywords: Air pollution, plant health, Thermal power plant, Emission of pollutant

1. Introduction

Reduced air pollution is a critical component of promoting long-term development. One of the worldwide and localized challenges is air pollution, mostly caused by energy generation and use. It has made a consistent contribution to global warming over time. In today's world, energy is fundamental in all parts of existence. It is critical for industrial production, transportation fuel and energy creation in thermal power plants. Also, electricity is widely known as a crucial instrument for almost all human activity. The exponential growth of the world's population is inextricably related due to increased energy consumption and environmental concerns. Who implies modern techniques to remediate exhaust emissions at power plants [1]. Using the finite volume approach with crude oil and gaseous fuel, Nasiriya power plant furnace, three-dimensional aerodynamic and thermal predictions at investigated liquid fuel combustion (crude oil or mazut) is solved using the discrete phase model to obtain acceptable temperature distribution, velocity, and other findings. According to the directions of a power plant industrial firm, the maximum temperature is 2293 K (at the furnace centre between the upper and lower burners), but it was discovered to be 2299.45 K in this job utilizing the Fluent software code. Also, according to this program, he found that when the firing was a gaseous fuel, the highest temperature was 2152 degrees, Kelvin, in practice and 2215 degrees Kelvin [2]. The Industrial Source Complex

Short–Term Model, Version 3 (ISCST3) was used to estimate atmospheric concentrations of sum suspended particles (TSP), Sulphur dioxide (SO2), and nitrogen dioxide (NO2) at seven Central Pollution Control Board-controlled watching sites (receptor locations) from July 2004 to June 2005. (CPCB). TSP and NO2 predictions from the ISCST3 model were accurate at all receptor locations. However, in SO2, the model predictions were only valid at two receptor sites [3]. For various fuels and boilers, standard emission factors for CO2, SO2, and NOx emissions from unregulated fossil-fuel power stations are widely available in the literature. Pure fuels with constant mean compositions are commonly referred to as these variables. However, the fuel mixtures can be recommended as BATs (best available techniques) to extend the boilers lives for environmental, economic, and technological reasons. Over eight years, this study used continuous emissions monitoring data to estimate relevant emissions parameters for a large 1400 MW coal-fired power station [4]. It had been looked into the impact of a shifting energy matrix for electricity production on air quality in a metropolitan region in a tropical, forested setting is examined. Manaus, Brazil's biggest city in the central Amazon Basin, has been transitioning its energy grid from fuel oil and diesel to natural gas for more than ten years [5]. He studied gaseous and liquid fuel combustion in the furnace of the Nasiriya power station and used the FLUENT algorithm to examine the mechanical properties with pollutants numerically [6,7]. In addition, directional vanes were added to the burner, which gave it a new look, several parameters influencing device operation were studied in detail. It is fundamentally dependent on the combustion of gases that have been released as a result of the spilt oil. The velocity distribution is a critical factor that provides the first forecast of whether or not a power turbine will be erected, calculates the high velocity at the burned gases cover collector when the chimney expands. A numerical study was performed using GAMBIT and FLUENT 6.3 [8]. Modelling of emitted gases in furnace chamber was used to investigate the parts of pollutants generated by fuel burn (liquid or gas) in oil and gas refineries. The case study was conducted at the Nasiriya power plant with various loads ranging from 70 to 210 MW. The finite volume approach was examined to anticipate pollutant fractions using the FLUENT. The case study was based on diagrammatically generated before being loaded into FLUENT and solved. NOx and SOx are two forms of waste organisms that have a noticeable impact on human health, as shown in table (1). The numerical analysis results used to calculate the contaminants in the chamber were as follows: When crude oil is burned, the amount of carbon dioxide released is larger than when gaseous fuel is used [9]. The CHP Zrenjanin has identified increased pollutant emissions throughout the fossil-fuel combustion process as a growing problem. Emissions on boiler B1 were measured over the winter period for five different burner working regimes by altering the fuel ratio amongst natural gas and fuel oil [10]. The software package Flex PDE is used to model the statistical dispersion of pollutants from power plants. As a result, using pollutant concentrations and ambient conditions, an estimate of the severity of air pollution can be made [11]. TPPs emissions would be reduced dramatically by moving to natural gas, while PM species, like BC, improved combustion and pollution control technology emissions would be reduced if this technology were implemented in small and medium-sized businesses. This study's pollutant data could be used in dispersion modeling for research on air quality control in Vietnam [12]. The estimation of monetary losses from PM2.5, SO2, NOx, NH3, and VOC exposure from electric power plants, oil and gas production, coal mining, and oil refineries were presented. In 2011, the expenses of these industries were 131 billion dollars (in 2000 dollars), with SO2 emissions from electricity generating accounting for the majority of these expenses [13]. According to studies recent, the addition of various nano additives to the fuel of emulsion of diesel (W / D) improves the properties of the fuel efficiency and reduce the levels of hydrocarbons (HC) and carbon monoxide of carbon (CO). In the W / D with 10% of water, they have selected nano additives of the following: oxide of aluminium, oxide of copper oxide to magnesium oxide of manganese and oxide of zinc (E10). To create one fuel of emulsion nano additive, E10Al2O3, E10CuO and E10MgO, MnO and ZnO were mixed in E10 to one dose of 50 ppm.All the fuels were tested for their properties and performance, as well as their emissions (oxides of nitrogen (NOx), particulates (PM), dioxide of carbon (CO2), CO and HC) to determine the best for each application. According to the results, the thermophysical properties of the fuel mixtures were enhanced, the heat transfer rate was increased, and the fuel mixtures were stabilized. Depending on the amount of nanofluid additives utilized, engine performance parameters improved and exhaust emissions decreased [14,15].

This paper aims to find standard solutions to air pollution issues produced by fossil fuel burning in thermal power plants, allowing for the prevention of lethal respiratory diseases. It also critically reviews the increased awareness of the significance of pollution control studies in places where thermal power plants were being built for electricity generation.

Table 1. Allowable Emission, 10^6 ton [9]

Source category	TSP	Sox	NOx	CO
Electric utilities				
Coal	2.6	17.5	5.6	
Oil	0.2	1.9	1.4	
Natural gas	0.1	0	0.9	
Total	2.8	19.4	7.9	0.3
Percentage of grand total	20	65	31	0.3
Other sources	11.0	10.4	17.8	112.2
Total	13.8	29.8	25.7	112.5

1.1. Emission study

An environmental assessment was carried out to examine the impact of this power system on the area surrounding the three-generation facility: the impact on native air quality was minimal. The ISC Gaussian Dispersion Model was used to conduct this environmental research [16]. In China's Shandong Province, a case study was led. Determine the efficacy of environmental standards in reducing pollutant emissions, and data were analyzed quantitatively. The findings revealed that each change to the substantial standard impacts pollutant discharge [17]. In addition, the impact of a wet FGD system on reducing SOx generated by the combustion of Mazut fuel in a boiler has been investigated. This research was both technical and economic [18].

A theoretical simulation of the contaminant's distribution and chemical interactions in the atmosphere during fuel burning at Ekibastuz SDPP-1 has been presented. A real-life example of a thermal power plant was used to model the dispersion of NO, NO2, CO, and products NO2, HNO3, CO2 waiting for chemical interaction with oxygen (Ekibastuz SDPP -1) [19]. The effects of different emission scenarios on NOx, SO2, PM total, and CO concentrations were simulated using the AERMOD air quality model. The first three scenarios revealed a significant reduction in NOx, SO2, and PM total emissions [20]. During the year 2014 in Kuwait, the author gives a case study on power plant emissions. Natural gas, gas oil, crude oil, and heavy fuel oil are all utilized in Kuwaiti power plants, with the type of fuel used fluctuating because of this fuel volatility. In this study, a fuel-switching method was simulated. When SO2 concentrations are projected to be high during ours, high-sulfur fuels will be replaced with natural gas. Emission estimates were combined with an air quality dispersion model to simulate the temporal and geographic variation of sulfur dioxide (SO2) and nitrogen dioxide (NO2) in Kuwait during one year. The findings reveal that averaging emissions and switching fuels reduces yearly maximum SO2 levels across the entire region see figure (1), [21]. Numerical simulations under real-world atmospheric conditions were used to investigate the mobility and dispersion of pollution emissions into the atmosphere. This set of equations was closed using the conventional k-epsilon and SST k-omega flow regimes, and the problem was solved using a system of Reynolds-averaged Navier–Stokes equations. The numerical solution of the test problem was also performed to ensure that the mathematical model and numerical methods were correct. The process of pollutant emissions from Ekibastuz SDPP chimneys and CO2 spread was defined using a validated mathematical model and numerical algorithm [22]. In 2020, data on SO2, NO2, and CO2 were collected from three monitoring stations in Al Hila district, Babylon governorate. In order to equate the concentrations of the critical contaminants in this city to the standards set by various international health organizations and provide a viable strategy for reducing air pollution's effects on the environment and human health [23].

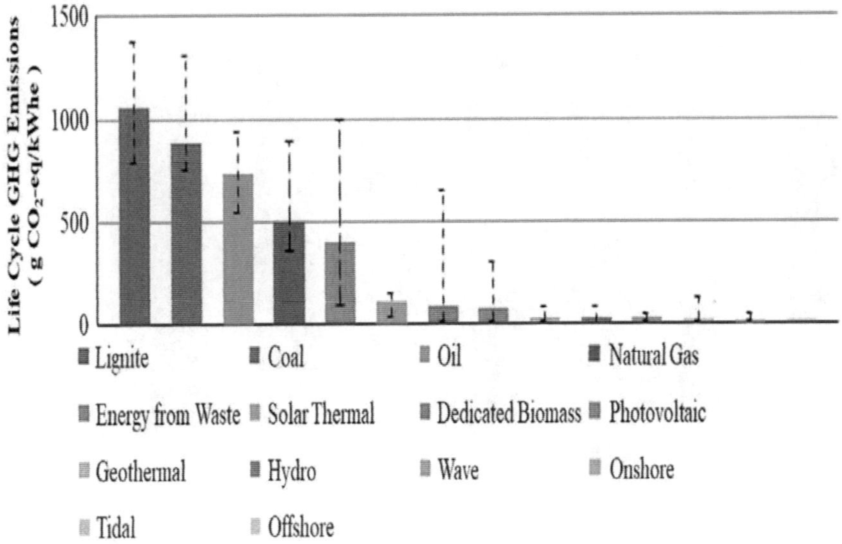

Figure 1. Estimates of GHG emissions over the life cycle of energy-generating [46]

2. Major air pollutants and their sources

The most significant pollutants in the atmosphere are CO2, SO2, NOx, and primary PM see figure (2), [24]. Carbon dioxide (CO2) is a colourless gas that is produced when carbohydrates are not entirely absorbed. When hydrocarbon fossil fuels, such as natural gas, gasoline, and coal, are used within the constraints of the atmosphere. Carbon dioxide is created when half of the oxygen that joins with the carbon atom is burnt [25,26]. Sulphur dioxide is a highly reactive colourless gas produced in power plants by fuel oil and coal [27]. Nitrogen dioxide is reddish-brown. Nitrogen oxides (NOx) are a group of gases produced by thermal power plants from burning fossil fuels. All of these pollutants are dangerous to people's health and the environment. Nitrogen in the air combines with oxygen [28,25]. Particulates (PM) are tiny particles having an aerodynamic diameter of less than or equal to 10 micrometres that form when fossil fuels are burned in coal-fired thermal power plants. Because of their small size, they can enter the airways deep within the lungs, where they might deposit and cause health problems [29,30].

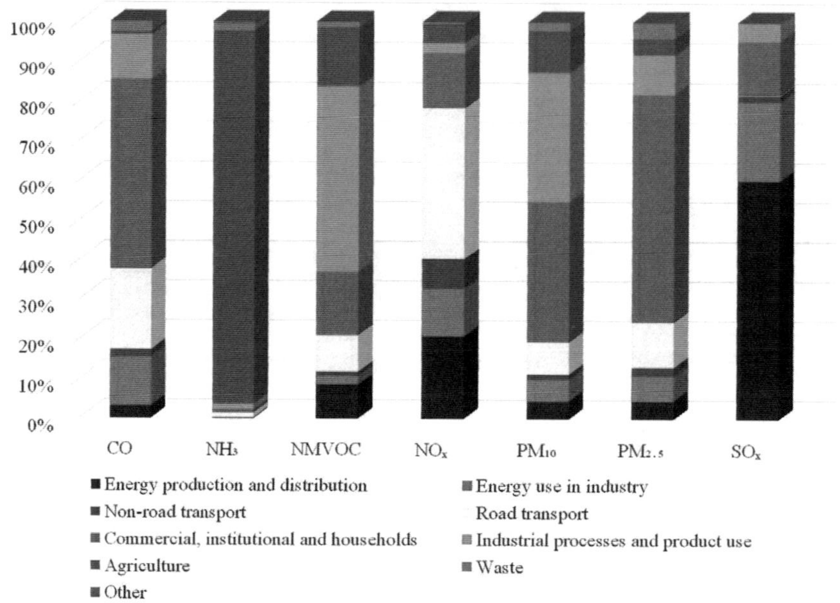

Figure 2. Emissions of central air pollutants by sector groupings in EA, based on EEA data [46]

2.1. Thermal Power Plant Emission Characteristics

Coal-fired power plants create flue gas that contains carbon dioxide and other pollutants, such as nitrogen, sulfur dioxide, fly ash (PM), and mercury, are all present in flue gas produced by the combustion of fossil fuels [12]. According to the study, emissions from coal-burning include acidic SO2, NO2 (sulphur and nitrous oxides) gases, water vapour, micron-sized particulate matter, and hazardous mercury, in addition to greenhouse gas emissions (CO, CO2 and N2O). Coal's calorific value, source, and amount all impact the emissions resulting [31]. In a power plant boiler, coal produces flue gas. It contains nitrogen, carbon dioxide, and water (H2O). It also transfers pollutants. Oxides such as sulfur oxides (SOx) and nitrogen oxides (NOx) can be present in the combustion process and the fuel used [32]. One of the most pressing environmental challenges in emerging countries is gas flaring. The Flaring of gases pollutes the air in the oil processing region and releases various pollutants into the environment, including NOx, CO, CO2, SOx, and total hydrocarbon.

2.2. Thermal Power Plant's Environmental Impact

According to [33], SO2 is an atmospheric contaminant produced mainly by electrical power stations that scorch high-sulfur coal. Therefore, SO2 scrubbers, low-sulfur coal, or gasification and sulfur recovery via regulations should be mandated for power stations, said the study's findings. Water vapour, sulfur dioxide, and other altitude compounds interact to generate sulfuric and nitric acids. Acids produced in clouds or fog are usually dissolved in the water droplets floating in the atmosphere. These acidic droplets are washed onto the ground like lemon juice from the air when it rains or snow. This is known as acid rain. According to [34], In 2015, it was predicted that 4.2 million people died due to exposure to ambient particulate matter worldwide. [35], Plant leaves are the initial receptors for PM contaminants in the surrounding atmosphere, and interacting with them damages the leaf surface.

PM contaminants have morphological, physiological, and biochemical effects on plants. [36], Dust on the foliar surface reduces plant growth by interfering with gaseous exchange through the stomata. [37], NO2 has the potential to irritate the lungs and cause pulmonary oedema. Epidemiologic studies show that higher concentrations irritate the eyes, nose, throat, headaches, dyspnea, chest discomfort, bronchi spasm, and pulmonary oedema. Coughs and breathing difficulties are caused by a high degree of NO2 exposure, which produces increased mucus secretion. Long-term NO2 exposure is linked to an increased risk of respiratory illnesses. People who already have asthma or emphysema are affected by SO2 in the air. It makes breathing more difficult for them. It also irritates the eyes, nose, and throat [38], SO2 degrades our plants and structures, making it impossible to observe long-distance things. A high dose of SO2 aggravates a cardiovascular illness that already exists. During mouth breathing, SO2 is inhaled more deeply into the lungs than during regular nasal breathing. SO2 is deposited on the lining of the airways. It degrades into sulfite or bisulfite, which is then disseminated throughout the body. It stimulates sensory receptors in the airways when it is in the form of sulfite or bisulfite. It breaks down into sulfite or bisulfite, which is then dispersed throughout the body. It activates sensory receptors in the airways and produces bronchial constriction in the form of sulfite or bisulfite. Because SO2 is so easily soluble in water, it causes acid rain and soil acidification. The presence of SO2 in water reduces the amount of dissolved oxygen in the water, resulting in aquatic life death. Skin redness and blisters are also caused by SO2 exposure. [39], A case study looked at how switching a Central Mexican power plant from fuel oil to natural gas lowered SO2 emissions. Power plant hourly fuel oil consumption data from a 2010 report on pollutants, greenhouse gases, organic and inorganic toxins were analyzed. Secondary SO2 products from power plants may be transported to other power plants under low pH conditions in the south of Mexico City, where high sulfate concentrations have been found. Secondary products may create acid rain in Mexico City's urban area, even though the environment around a power station is toxic. Mexico City's northern suburbs are home to an oil refinery that produces most of the country's SO2 emissions. Significant reductions in sulfur dioxide (SO2) emissions from facilities could result in decreased acid rain levels in the immediate vicinity of populated areas and should be made mandatory to meet annual emission limits and health standards. In order to meet annual emission limitations and health standards, switching to cleaner fuels should be mandated in critical places. This might reduce acid rain levels in neighboring populated areas by as much as 40%.

3. Thermal Power Plant Exhaust Gas Treatment

3.1. Carbon Dioxide Treatment

The quantity of CO2 emitted per MW of power generated by fossil fuels fluctuates substantially due to generation efficiency, fuel options, and plant age changes. The average CO2 emissions per MW have been gradually dropping because of low-carbon fuels such as natural gas and improvements in energy conversion efficiency. After burning, the gas is detached from flue gases in the post burning capture of the CO2 process. CO2 sequestration from power plants can only begin after the gas has been caught in its most pure form. CO2 sequestration is the process of storing or using CO2 in a method that keeps it out of the atmosphere. A lesser fraction of the CO2 gathered could create chemicals, while massive amounts may be used to boost oil and gas output. CO2 storage could take several forms, including underground storage and ocean sequestration. CO2 removal from a gas stream can be accomplished in various ways, including amine absorption, various adsorption techniques, membranes, and upon. 85-95 per cent of CO2 in the fuel may be recovered during the capture procedure. Solid compounds that react with the gas in a solid state can also be used to capture CO2. These CO2capture techniques consume much energy, lowering electricity generation efficiency by up to 40% and increasing net power output by 40%. CO2 capture after combustion is commonly recognized as one of the essential strategic alternatives for reducing CO2 emissions from coal-fired power plants [40,41,42].

3.2. SOx and NOx Treatment

Due to environmental concerns, several coal-fired power plants have constructed denitrification (DeNOx), and desulphurization (DeSOx) plants to reduce nitrogen oxide (NOx) and sulfur oxide (SOx) emissions. Wet flue gas desulfurization is one of the most used ways for lowering SOx emissions (WFGD). However, the principal approaches for lowering NOx in the combustion chamber only have a 35% success rate. Moreover, it is not enough to comply with environmental laws. Secondary NOx removal methods include Non-Catalytic Decrease (SCR) and SCR (SCR). These methods, which are outside of the boiler combustion zone, are more promising. Secondary measures, such as SNCR and SCR, can achieve efficiencies of 50% and 95%, respectively, [31,43,44]. Coal quality and the plant's environmental management policy are two critical determinants of flue gas treatment. Coal-fired power plants in some countries are required to release DeNOx and DeSOx.

According to the study, while the continent may differ from other countries, Australian coal has a low sulphur level. Therefore the situation may differ in other countries. In addition, the concentration of SOx produced by coal burning in Australia is relatively low. As a result, there were no harsh regulatory constraints. As a result, in the past, DeNOx or DeSOx devices to clean up or treat NOx and SOx in flue gas streams in Australian coal plants were not necessary [31]. A study was presented on the use of methods to reduce polluting nitrogen oxides and sulfur oxides in the Nasiriya power plant, which uses oil and gas fuels, by injecting water and recycling flue gases to drive the high-temperature gradient from the front wall while lowering nitrogen oxides and sulfur oxides concentrations [2].

3.3. Particulate Matter Treatment

The categorization and collection effectiveness of particle emission control systems (PECDs) were evaluated. Particulate acceptable emission efficiency and size allocation were measured using anthracite and bituminous charcoal at four types of coal-fired power units. The mass concentrations of PM1, PM2.5, and PM10 were measured in the range of aerodynamic diameters from 0.03m to 10m. Researchers showed that electrostatic precipitators were 99.0-99.89 per cent effective at collecting PM and PM10, respectively. PM and PM10 bag-stock collection rates were 99.94% and 99.76%, respectively. Group efficiency was 90.8-98.6 per cent for ESP and 99.54 per cent for bag-stock with particle sizes that ranged from 0.1 to 1.1 millimetres, respectively, according to the study [45].

4. Results and Discussion

The findings of this study yielded predictions that corresponded to the literature assessment of gas flow and aerodynamics inside the furnace and every other temperature, species, and combustion product concentration. This experiment was conducted using liquid fuel (crude oil) at full load (210MW), with flue gas recirculation switched off, because using flue gases to minimize NOx at this time was not advantageous [6]. Recent research shows that adding nano-additives to diesel emulsion fuel (W/D) can improve fuel characteristics and performance while lowering hydrocarbon (HC) and carbon monoxide (CO) emissions. These nano-additives are utilized in W/D with 10% water (ZnO) [15]. Long-term energy supply security, fossil fuel resource exhaustion, global warming, and other energy-related environmental issues, as well as secure energy supply and rising fuel prices, are all key concerns. Renewable energy sources may readily meet current and future energy demands while inflicting minimum harm to the global environment.

5. Conclusions

This study focuses on air pollution, and it examines the primary sources and types of air pollutants, highlighting the limitations and challenges. Air pollutants are mainly from energy generation. It is crucial to use the recirculation of the flue gases system to push the high-temperature gradient from the front wall and reduce the concentration of NO pollution. The NO concentration is highest at 70MW and gradually decreases as the load increases, reaching a minimum value at maximum load (210MW). The flue gases recirculation system is operated at 70MW while being turned off at 200-210MW. With the high-temperature gradient and low velocity, the maximum amount of NO pollutant for crude oil is concentrated at the line of symmetry of the furnace (X=0.5) close to the front wall of the furnace, this is also the rationale for operating the flue gases recirculation at this position. The chimney's height reduces the concentration of pollution that is badly affecting the life in the earth due to the most plumes going far distances and the percentage of concentration becomes very small. High smoke velocity from the chimney gives the plumes a tremendous tall and distance to reduce the concentration because the high velocity should produce the excellent dispersion of the plumes. The best smoke velocity is more than 1.5 air velocity because the high velocity of air causes downwash for the pollutants, and the minimal velocity cannot remove the plumes for far distance.

The evaluation of thermal power plants' handling of exhaust gases was completed effectively. The dangers posed by the contaminants and their emissions were detailed in great detail. These emissions can be reduced by employing best practices that include CO2 capture to desulfurization and denitrification to PM collection. The majority of the research examined adheres to established methods and is consistent with their conclusions. The preceding sections discussed nano-additives in detail, from their manufacturing to their usage in diesel and biodiesel fuels. The use of nanoparticles in liquid fuels has many potential benefits, including increased performance and reduced emissions. In over 90% of the literature reviewed, unburned hydrocarbons (UBHC), carbon monoxide (CO), and particulate matter (PM) emissions were reduced (PM). Metal-based (Mn, Ni, Mg, Co) additives and a few oxygenated additives reduce the biodiesel blend's viscosity, density, and flashpoint. Recent research has focused on oxygenated additives because of their low cost and availability of synthesis machinery. Because the temperature of the combustion chamber increases as engine load increases, the oxygenated additions reduce CO emissions.

Additionally, the oxygenated chemicals have a natural cooling effect, which results in a more excellent cylinder. The fuel properties, engine performance, and emissions of D2, E10, E10Al2O3, E10CuO, E10MgO, E10MnO, and E10ZnO were investigated. The findings allow for certain inferences to be drawn. There was no discernible variation in the viscosity of the nano-additives emulsion fuels as compared to E10. The findings of this study can be utilized to determine the benefits of reducing atmospheric pollution and thereby eradicating acid rain. Additionally, CO2 collection contributes to addressing the grave problem of global warming, and delivering high-quality air may help prevent severe respiratory illnesses. However, additional work is necessary to optimize the bag-house collection system and, ultimately, the disposal technique.

References

[1] Asif, M., & Muneer, T. (2007). Energy supply, its demand and security issues for developed and emerging economies. Renewable and sustainable energy reviews, 11(7), 1388-1413.

[2] Hannun, R. M. (2009). Three Dimensional Aerodynamic and Temperature Predictions for Nassiriya Power Plant Boiler Furnace (Doctoral dissertation, Ph. D. Thesis, College of Engineering, Basrah University, Iraq).

[3] Kansal, A., Khare, M., & Sharma, C. S. (2011). Air quality modelling study to analyse the impact of the World Bank emission guidelines for thermal power plants in Delhi. Atmospheric Pollution Research, 2(1), 99-105.

[4] Dios, M., Souto, J. A., & Casares, J. J. (2013). Experimental development of CO2, SO2 and NOx emission factors for mixed lignite and subbituminous coal-fired power plant. Energy, 53, 40-51.

[5] Medeiros, A. S., Calderaro, G., Guimarães, P. C., Magalhaes, M. R., Morais, M. V., Rafee, S. A., ... & Souza, R. A. (2017). Power plant fuel switching and air quality in a tropical, forested environment. Atmospheric Chemistry and Physics, 17(14), 8987-8998.

[6] Hannun, R. M., Najim, S. E., Rishack, Q. A., & Syred, N. (2011). 3-D. Aerodynamics And Thermal Prediction For Nassiriya Power Plant Furnace By Using Crude Oil Fuel‖. International Journal of Advanced in Science and Technology, 3(4).

[7] Hannun, R. M., Najim, S. E., Rishack, Q. A., & Syred, N. (2014). 3-D. Aerodynamics and Thermal Prediction For Nassiriya Power Plant Furnace By Using Gaseous Fuel. IOSR Journal of Engineering, 4, 12-26.

[8] Hannun, R. M. (2013). New Base and Burner Design for Utilizing the Compound Gases with Spilled Oil at Using Power Chimney Techniques. The University of Thi-Qar Journal for Engineering Sciences, 4(2), 14-29.

[9] Hannun, R. M., Najim, S. E., Rishack, Q. A., & Syred, N. (2013). Modelling of Pollutants Prediction from Fuel Burning in Oil and Gas Refineries. Engineering and Technology Journal, 31(18 Part (A) Engineering).

[10] Nakomcic-Smaragdakis, B., Cepic, Z., Cepic, M., & Stajic, T. (2014). Data analysis of the flue gas emissions in the thermal-power plant firing fuel oil and natural gas. International Journal of Environmental Science and Technology, 11(2), 269-280.

[11] Mărunţălu, O., Lăzăroiu, G., & Bondrea, D. A. (2015). A mathematical model for air pollutants dispersion emitted by fuel combustion. UPB Sci. Bull., Series D, 77 (4), 229-236.

[12] Huy, L. N., & Oanh, N. T. K. (2017). Assessment of national emissions of air pollutants and climate forcers from thermal power plants and industrial activities in Vietnam. Atmospheric Pollution Research, 8(3), 503-513

[13] Jaramillo, P., & Muller, N. Z. (2016). Air pollution emissions and damages from energy production in the US: 2002–2011. Energy Policy, 90, 202-211.

[14] Soudagar, M. E. M., Nik-Ghazali, N. N., Kalam, M. A., Badruddin, I. A., Banapurmath, N. R., & Akram, N. (2018). The effect of nano-additives in diesel-biodiesel fuel blends: A comprehensive review on stability, engine performance and emission characteristics. Energy Conversion and Management, 178, 146-177.

[15] Hasannuddin, A. K., Yahya, W. J., Sarah, S., Ithnin, A. M., Syahrullail, S., Sidik, N. A. C., ... & Ramlan, N. A. (2018). Nano-additives incorporated water in diesel emulsion fuel: fuel properties, performance and emission characteristics assessment. Energy Conversion and Management, 169, 291-314.

[16] Iodice, P., d'Accadia, M. D., Abagnale, C., & Cardone, M. (2016). Energy, economic and environmental performance appraisal of a trigeneration power plant for a new district: Advantages of using a renewable fuel. Applied Thermal Engineering, 95, 330-338.

[17] Yuan, X., Zhang, M., Wang, Q., Wang, Y., & Zuo, J. (2017). Evolution analysis of environmental standards: Effectiveness on air pollutant emissions reduction. Journal of Cleaner Production, 149, 511-520.

[18] Kouravand, S., & Kermani, A. M. (2018). Investigation on Influence of Wet FGD to Reduction of SOx from the Flue Gases due to Combustion of Mazut in Boilers. Russian Agricultural Sciences, 44(4), 385-391.

[19] Issakhov, A., Alimbek, A., & Issakhov, A. (2020). A numerical study for the assessment of air pollutant dispersion with chemical reactions from a thermal power plant. Engineering Applications of Computational Fluid Mechanics, 14(1), 1035-1061.

[20] Alhajeri, N. S., Dannoun, M., Alrashed, A., & Aly, A. Z. (2019). Environmental and economic impacts of increased utilization of natural gas in the electric power generation sector: Evaluating the benefits and trade-offs of fuel switching. Journal of Natural Gas Science and Engineering, 71, 102969.
[21] Al-Fadhli, F. M., Alhajeri, N. S., Aly, A. Z., & Allen, D. T. (2019). The impact of power plant emission variability and fuel switching on the air quality of Kuwait. Science of The Total Environment, 672, 593-603.
[22] Issakhov, A., & Mashenkova, A. (2019). Numerical study for the assessment of pollutant dispersion from a thermal power plant under the different temperature regimes. International Journal of Environmental Science and Technology, 16(10), 6089-6112.
[23] Al-Salman, A. A., & TaghiEbadi, R. M. (2021). Mitigation Strategies for Reducing of Air Pollutant, Case Study: Al-Hilla City-Babylon Province/Iraq. Annals of the Romanian Society for Cell Biology, 7460-7469.
[24] Schneider, D. R., & Bogdan, Ž. (2007). Effect of heavy fuel oil/natural gas co-combustion on pollutant generation in retrofitted power plant. Applied thermal engineering, 27(11-12), 1944-1950.
[25] Nazari, S., Shahhoseini, O., Sohrabi-Kashani, A., Davari, S., Paydar, R., & Delavar-Moghadam, Z. (2010). Experimental determination and analysis of CO2, SO2 and NOx emission factors in Iran's thermal power plants. Energy, 35(7), 2992-2998.
[26] Rubin, E. S., Chen, C., & Rao, A. B. (2007). Cost and performance of fossil fuel power plants with CO2 capture and storage. Energy Policy, 35(9), 4444-4454.
[27] Ramadan, A. A., Al-Sudairawi, M., Alhajraf, S., & Khan, A. R. (2008). Total SO2 emissions from power stations and evaluate their impact in Kuwait using a Gaussian plume dispersion model. American Journal of Environmental Sciences, 4(1), 1-12.
[28] Saini, D. K., Garg, S. K., & Kumar, M. (2019). MAJOR AIR POLLUTANTS AND THEIR EFFECTS ON PLANT AND HUMAN HEALTH: A REVIEW. Plant Archives, 19(2), 3273-3278.
[29] Contini, D., Cesari, D., Conte, M., & Donateo, A. (2016). Application of PMF and CMB receptor models for evaluating the contribution of a large coal-fired power plant to PM10 concentrations. Science of the Total Environment, 560, 131-140.
[30] Shin, D., Woo, C. G., Hong, K. J., Kim, H. J., Kim, Y. J., Han, B., ... & Chun, S. N. (2020). Continuous measurement of PM10 and PM2. 5 concentration in coal-fired power plant stacks using a newly developed diluter and optical particle counter. Fuel, 269, 117445.
[31] Rasul, M. (Ed.). (2013). Thermal Power Plants: Advanced Applications. BoD–Books on Demand.
[32] Kahforoshan, D., Fatehifar, E., Babalou, A. A., Ebrahimin, A. R., Elkamel, A., & Soltanmohammadzadeh, J. S. (2008, September). Modelling and evaluation of air pollution from a gaseous flare in an oil and gas processing area. In WSEAS Conferences, Santander (pp. 180-186).
[33] Najjar, Y. S. (2011). Gaseous pollutants formation and their harmful effects on health and the environment. Innovative energy policies, 1, 1-9.
[34] Howard, D. B., Thé, J., Soria, R., Fann, N., Schaeffer, R., & Saphores, J. D. M. (2019). Health benefits and control costs of tightening particulate matter emissions standards for coal power plants-The case of Northeast Brazil. Environment international, 124, 420-430.
[35] Rai, P. K. (2016). Impacts of particulate matter pollution on plants: Implications for environmental biomonitoring. Ecotoxicology and environmental safety, 129, 120-136.
[36] Sett, R. (2017). Responses in plants exposed to dust pollution. Horticulture International Journal, 1(2), 53-56.
[37] Kim, D., Chen, Z., Zhou, L. F., & Huang, S. X. (2018). Air pollutants and early origins of respiratory diseases. Chronic diseases and translational medicine, 4(2), 75-94.
[38] Chen, T. M., Kuschner, W. G., Gokhale, J., & Shofer, S. (2007). Outdoor air pollution: nitrogen dioxide, sulfur dioxide, and carbon monoxide health effects. The American journal of the medical sciences, 333(4), 249-256.
[39] Sosa E, R., Vega, E., Wellens, A., Jaimes, M., Fuentes G, G., Granados H, E., ... & Mateos D, E. (2020). Reduction of atmospheric emissions due to switching from fuel oil to natural gas at a power plant in a critical area in Central Mexico. Journal of the Air & Waste Management Association, 70(10), 1043-1059.
[40] Liu, L., Wang, S., Niu, H., & Gao, S. (2018). Process and integration optimization of post-combustion CO2 capture system in a coal power plant. Energy Procedia, 154, 86-93.
[41] Marion, J., Nsakala, N. Y., Griffin, T., & Bill, A. (2001, May). Controlling power plant CO2 emissions: a long-range view. In 1st National Conference on Carbon Sequestration, Washington, DC (Vol. 15, p. 17).
[42] Sani, F. M., & Oyelaran, O. A. (2019). Exhaust Gas Treatment in Thermal Power Plants: A Review. International Journal of Advances in Scientific Research and Engineering, 5(11), 227-233.
[43] Center, J. C. E. (2007). Clean Coal Technologies in Japan: Technology Innovation in the Coal Industry. JCOAL, Tokyo, Japan.
[44] Madejski, P. (2018). Introductory chapter: new trends and recent developments for thermal power plants. In Thermal Power Plants-New Trends and Recent Developments. IntechOpen.
[45] Bandyopadhyay, A., & Mandal, P. K. (2015). Investigation on multiple benefits of wet ESP for emission control in Indian coal-fired thermal power plant. Clean Technologies and Environmental Policy, 17(2), 421-431.
[46] Wang, X. C., Klemeš, J. J., Dong, X., Fan, W., Xu, Z., Wang, Y., & Varbanov, P. S. (2019). Air pollution terrain nexus: A review considering energy generation and consumption. Renewable and Sustainable Energy Reviews, 105, 71-85.

IOP Conf. Series: Earth and Environmental Science 1002 (2022) 012009 doi:10.1088/1755-1315/1002/1/012009

Water management scheme to restore and sustain the Marshes and Shatt al-Arab, Southern Iraq.

[1] Khayyun Rahi, and Ali Abdulkhabeer Ali [2]

*[1] Independent Scholar (retired Assoc. Professor), Tulsa, Oklahoma, USA,

**[2] Marshes Research Center, University of Thi-Qar, Nasiriyah, Iraq

* corresponding author: khayyun.rahi@okstate.edu

** aranru79@utq.edu.iq

Abstract

Attempts to restore the Marshes of Southern Iraq (re-marshification) have started in the summer of 2003. After about 18 years, very little has been achieved in the restoration process. The inundated areas fluctuate on an annual basis. The water-covered areas increase during wet years and shrink during dry years. One component of the re-marshification is missing; that is a plan based on actual water flow data and supplies. As a result, the vast area of the original marshes is suffering from saline water and ecological deterioration. In this paper, sustained water supplies are specified to restore an area of about 4440 km2, about 55.5% of the marshes area of 1973, and reclaim the salinity problem of Shatt al-Arab. The basic components of the water supplies scheme come from two sources: the first is the environmental flow from the Tigris, the Euphrates, the Karkheh, and the Karun rivers, and the second is the seasonal flows generated along the eastern borderline of Iraq (i.e., The Iraq-Iran Borderline). The proposed scheme requires some engineering works to convey the water flow to the intended Marshes area and isolate Shatt al-Arab from the marshes hydraulically.

Key words: Water management, Sustainable environment, Marshes of Iraq, Shatt al- Arab.

1. Introduction

Iraq relies on the twin rivers, the Tigris and the Euphrates, as the sole source for its water supply. The Euphrates and the Tigris are both international rivers that originated in Turkey. The Euphrates starts in southeastern Turkey and runs southward for about 450 km, then enters Syria flowing southeast for about 680 km [1,2]. The Euphrates enters northern Iraq via al-Qaim.It flows south and southeast across Iraq for approximately 1870 kilometers till it meets the Tigris near al-Qurna (southern Iraq) [3].The Tigris River is approximately 1900 kilometers long, with 400 kilometers in Turkey and 32 kilometers in Syria and Iraq.[4,5].The river travels southeasterly till it reaches al Qurna, where it joins the Euphrates. The confluence of the Tigris and Euphrates rivers results in the formation of the Shatt al Arab River. Shatt al-Arab runs approximately 192 kilometers southward into the Persian (aka Arabian) Gulf (hereinafter referred to as the Gulf) [6].Two Iranian rivers, the Karkheh and the Karun, form the other two tributaries of Shatt al-Arab.Since the mid-1970s, the Euphrates and Tigris rivers have been flowing into Iraq less due to dams and reservoirs built upstream, primarily by Turkey's Southeast Anatolia Project (Guneydogu Anadolu Projesi; GAP) [3,7,8].GAP was initiated about 1977 [9].

The GAP is an irrigation and agricultural project that relies on the Euphrates and Tigris Rivers for water. The project includes at least 30 dams and numerous other hydraulic structures, all of which are located in Turkey [2,7,10].The GAP dams on the Euphrates have a gross storage capacity five times that of the river's annual flow and twice that of the Tigris [7].The GAP and other dams in Syria resulted in excessive regulation of the Euphrates and Tigris rivers. The two rivers' uncoordinated over-regulation processes reduce runoff and increase the salinity of the inflow into Iraq [3,5].As a result, Iraq has been suffering from a water supply crisis for the past three decades. The Iraqi water crisis is evident in the lower reaches of the two rivers. Water scarcity affected the four southern governorates (Samawah, Nasiriyah, Basra, and Amara), the southern marshes, and the Shatt al-Arab River [6].The Shatt al-Arab was a freshwater river, except in its lower reaches, where the water was classified as brackish [11]. According to the same author, fresh river water is mixed with salty seawater, causing salinity to rise to around 2000 ppm at Seeba and downstream. "Developments on the Karun and Karkheh rivers in Iran may allow serious intrusion of seawater into the Shatt-el-Arab, with a consequent deterioration of its riparian date gardens, unless fairly substantial contributions from the Tigris and Euphrates," a consultant engineering firm stated in 1963 [12].The salinity of Shatt al-Arab has increased dramatically over the last five decades [6,8].In 2009, water salinity measured by the Iraqi Ministry of Water Resources' National Center for Water Resources Management (IMoWR) was much higher than the permissible limit (18500 ppm) [13]. The marshes were devastated late in the last century as a result of the combined impact of intentional drainage works (implemented in the late 1980s and early 1990s) and upstream damming (the Turkish GAP and dams of Syria and Iraq).The Central and Al Hammar marshlands have dried completely, with 97 percent and 94 percent of their land cover transformed into bare land, respectively, while less than a third of the al-Hawizeh marsh remains flooded [9].Iraq began restoring the marshes in 2003, following the fall of the previous regime. To replenish the marshes, water has been diverted to the Tigris and Euphrates. There have been numerous studies conducted on the quality of water [14,15]. The Iraqi Ministry of Water Resources set a goal of restoring 75 percent of the marshland area from 1973. Nonetheless, efforts to achieve the ministry's objectives have been futile.The re-flooded area of the marshes has fluctuated between 20% (in 2009) and 50% over the last 18 years (in 2006).The marshes are being flooded in an unsustainable manner due to a general water shortage and the lack of a long-term strategy to maintain a constant flow to the marshes. A water management plan that allocates continuous water flow to restore the marshes and Shatt al-Arab is outlined in this paper.
The proposed flow originates from two sources: the environmental flow of the Euphrates, Tigris, Karkheh, and Karun, and the uncontrolled sheet flow along the Iraq-Iran borderline.

2. The study area and the restoration process

The study region encompasses southern Iraq's marshes, including al-Hawizeh, the Central and al-Hammar wetlands, as well as the Shatt al-Arab River (Fig, 1a). The original wetlands cover an area of between 8000 and 20000 km2 [7,9].The overall area addressed in this paper is 8000 km2, which corresponds to the area that was flooded in 1973 [7].Three governorates contain marshes.According to the 1973 flooded area, 41% of the marshes are located in Missan Governorate, 40% in Thi-Qar Governorate, and the remainder in Basra Governorate (Fig 1b).The marshes are an integral part of Iraq's water resources system.The Central and al-Hammar marshes have nearly totally dried out and converted into barren ground and salt crusts as a result of prior regime desiccation efforts. Al-Hawizeh wetland has been reduced to less than a third of its former size. Attempts to rehabilitate the wetlands began in 2003 but have met with little success (Fig. 2). Each 1000 km^2 of marshland requires 77 m3/sec of water to be restored and remain inundated throughout the year [16]. The total flow required to reclaim the wetlands area from 1973 is approximately 616 m3/sec. The Shatt al-Arab River originates at al-Qurna, at the confluence of the Euphrates and Tigris rivers (Fig. 1a). From there, the Shatt runs south-east toward the Gulf, passing through the eastern section of the Basra governorate. Historically, the Shatt connected lower Mesopotamia's wetlands to the Gulf.

Numerous streams branch off from Shatt al-Arab in and around Basra.The stream network is regarded as the fundamental irrigation and drainage infrastructure for agricultural areas adjacent to rivers [17].Shatt al-Basra is another surface water body that may contribute to southern Iraq's hydrological system. Shatt al-Basra transmits the flow of the Tigris and Euphrates Main Outfall Drain (MOD) to the Gulf (Fig. 1a).

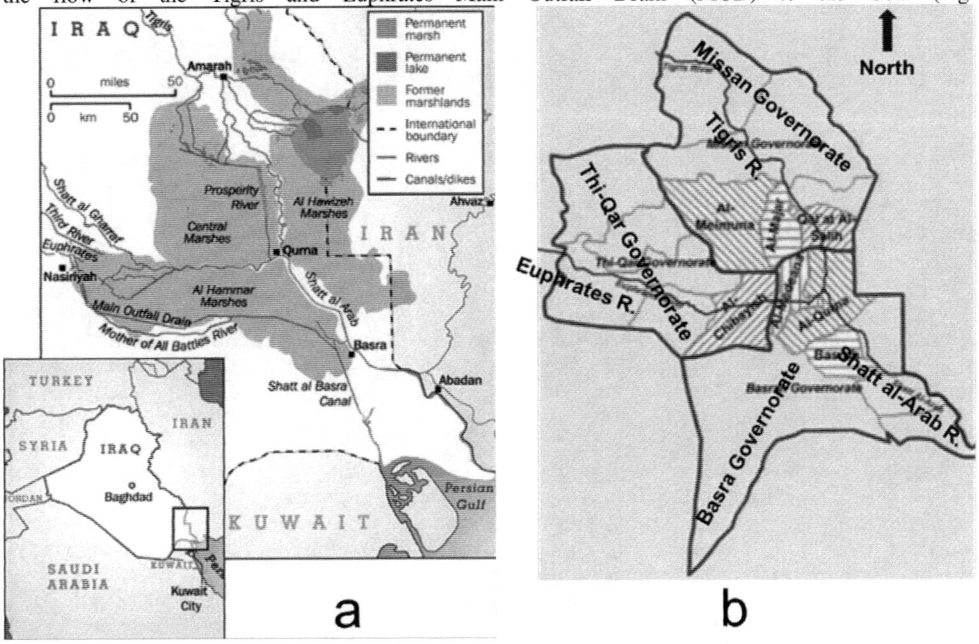

Fig. 1. (a) Location map for the marshes and Shatt al-Arab. (b) Marshes and respective governorates. (source: https://satoyama-initiative.org/old/iraq-traditional-agriculture-by-marsh-arabs-in-the-southern-marshes/) (accessed on July 2nd, 2021)

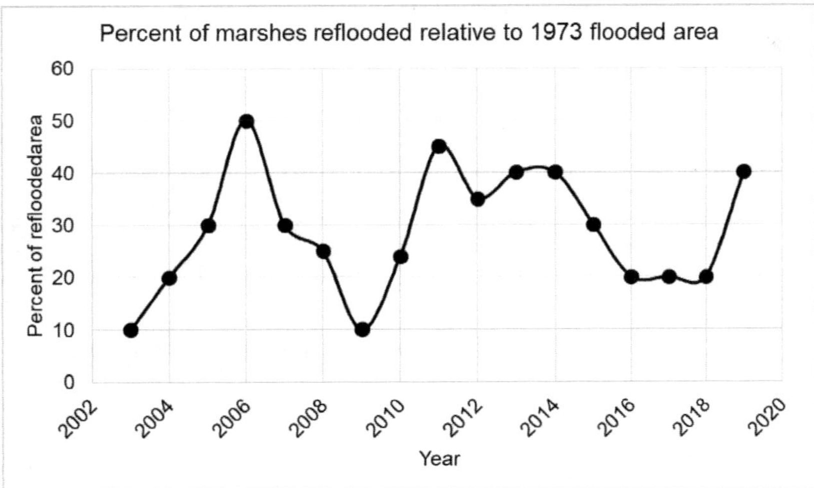

Fig. 2. an estimate of the marshland's reflooded area in relation to the marsh's in a flooded area in 1973.

3. Methodology and analysis

The proposed water management scheme comprises two flows elements: the environmental flows of the area rivers and the uncontrolled small wadies flow along the Iran-Iraq borderline.

a. The environmental flows

A river is an ecosystem that requires a continuous flow of freshwater to enable it to serve humankind and nature properly. The continuous flow is, in fact, the environmental flow. Environmental flows (e-flows) are not new but evolved over time [18].

The term (e-flows) was initiated in the Brisbane Declaration as "environmental flows describe the quantity, quality and timing of water flows required to sustain freshwater ecosystems and the human livelihoods and well-being that depend on these ecosystems" [19]. In the legal context, the term e-flow can vary substantially. European Union Water Framework Directive (WFD) uses the term "ecological flows" instead of e-flows. The ecological flows involve the hydrological regime that achieves a good ecological status of water bodies, which is the environmental objective of WFD [20]. Thus, the term e-flows are a major shift from the classical concept of minimum water requirement for a stream to a more comprehensive approach to cover the river system dynamics. In the literature, several definitions are published describing the e-flows; some are quoted below. Managing freshwater resources in a way that reflects the natural flow of things is referred to as "Environmental Flows."

As defined by scientific research, Environmental Flows represent the quantity, quality, and timing of water that is required to support a river and its related fauna [21].As a tool, use the flow of nature:

It is the ability to control river flows in accordance with the demands of people, agriculture, industry, energy and the environment within the constraints of available supply that environmental flows provide. The water-energy-food nexus can be effectively managed through the use of environmental flow.[22].

The United Nations Environmental Program (UNEP) defines environmental flow as the quantity, frequency, timing, and quality of water and sediment flows required to support freshwater and estuarine ecosystems, as well as human livelihoods and well-being that rely on these ecosystems [23]. Smakhtin and Anputhas [24] simplify the concept of the environmental flows as 'environmental demand' similarly to crop water requirements, industrial or domestic water demand.

The e-flows, as far as this paper is concerned, are the classic e-flows or the minimum instream flows (MIF) to sustain a healthy environment for the river system and preserve its biodiversity. The rivers that contribute to the water resources of the marshes and Shatt al-Arab are the Euphrates, the Tigris, the Karkheh, and the Karun. MIF for these rivers is calculated based on hydrological methods.

The MIF for the Euphrates River was calculated based on discharge-salinity correlation and salinity threshold [3,25]. The equation for the Euphrates MIF is [3]

$$S = 8{,}145 \text{ x } Q^{-.4045} \quad (1)$$

where S is the salinity reflected by TDS (ppm) and Q is the discharge (m3/s). The salinity in Equation (1) is fixed at 1000 ppm.

The MIF for the Tigris is calculated using two equations and two locations: Kut and Baghdad [5]. The two equations are:

$$\text{TDS} = 5539 \text{ x } Q^{-.325} \quad (2)$$

where TDS is the total dissolved salts in ppm (fixed at 1000 ppm) and Q is the discharge (m3/s).

$$\text{TDS} = 4123 \text{ x } Q^{-.274} \quad (3)$$

The variables as defined in Equation (2)

For the remaining rivers that originated in Iran, the e-flows are calculated using a hydrological method that allocated 10% of the mean annual stream flow as MIF (hereafter called the 10% method). The method is implemented in the USA and Turkey among other countries [20].

b. The uncontrolled small wadies flow along the Iran-Iraq borderline

Along with the e-flows, the water management plan presented by this paper includes uncontrolled flows mostly coming from Iran toward the Tigris River basin. The border between Iran and Iraq is marked by a large number of large outlying valleys and rivers, as well as several smaller ones. These valleys begin in western Iran and flow eastward into Iraq, with the bulk of them beginning in Iran.

The catchment areas of the valleys cover an area of approximately 20,000 km^2 in both Iraq and Iran [26].

The Iraqi portion of the watershed is mostly flat and gently sloping, with heights varying from 400 m above mean sea level (mainland latitudes) along the international perimeter to around 25 m above mean sea level (mainland latitudes) near the region's southeast end.

Among the numerous valleys, this article will concentrate on the two southernmost rivers, the al-Teeb and the Dwaireeg (Fig. 3), as well as a number of valleys to the northwest of the al-Teeb.

In their study of the flow of the al-Teeb and Dwaireeg rivers in eastern Iraq, Rahi et al. [26] came to the conclusion that the following equation best approximates the flow of the rivers: $Q_{mean} = 5.66\, A^{0.673} E^{-0.605} P^{1.03}$ (4)

where A is the catchment areas, $mile^2$; E is the mean catchment elevation in thousands of feet; P is the mean annual precipitation, inches; The estimated annual flow of both peripheral rivers is 100 m3/sec., considering this flow is seasonal (rainy season) and not a year around.

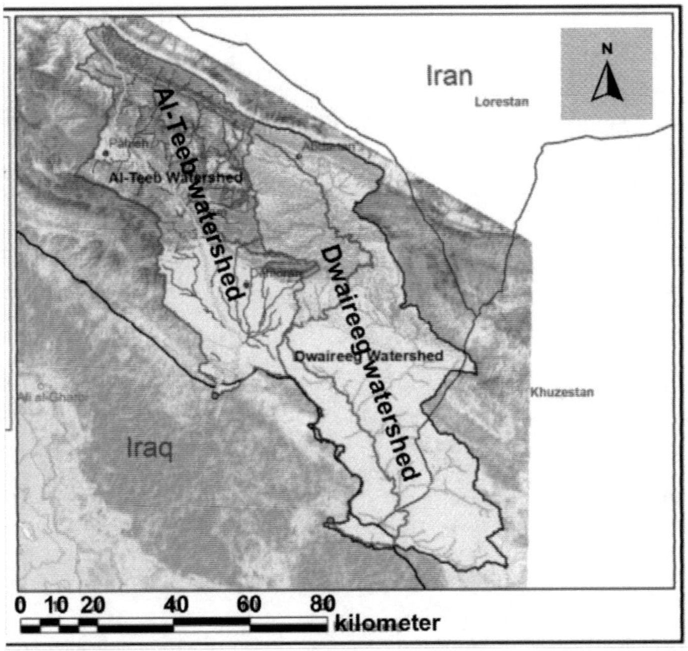

Fig. 3. al-Teeb and Dwaireeg rivers watersheds (adapted from [24])

4. Results

The e-flows of the Euphrates resulted using Equation (1) is 178 m3/sec. If the 10% method is employed, the e-flows becomes 105 m3/sec. According to the accessible literature [27], Keban gauge station in Turkey measured 261 (ppm) of total dissolved solids (TDS). Therefore, if the 10% approach is adapted, the water should be of 260 ppm salinity or less.

The Tigris e-flows, as calculated by Equations (2 and 3), and is 185 m^3/sec [5]. If the e-flows are calculated based on the 10% method, the e-flows are 154 m^3/sec. Due to the high salinity of the waters of the lower parts of Tigris and the Euphrates, the paper will adopt the e-flows of 185 m^3/sec and 178 m^3/sec for the two rivers, respectively.

E-flows of al-Karkheh and the Karun rivers are calculated based on the 10% method. The mean annual flows of al-Karkheh and the Karun are 5.8 billion cubic meters (bcm) and 24.5 bcm, respectively [8]. Accordingly, the e-flows for the Karkheh and the Karun are 19 (m^3/s) and 78 (m^3/s), respectively.

The uncontrolled flow along the Iran-Iraq borderline includes two components: al-Teeb and Dwaireeg rivers and the numerous valleys to the northwest of al-Teeb. Flow estimation is available for the two rivers but not for the northwestern valleys. Finding an estimate of the valleys' flow is beyond the scope of this paper. Therefore, the flow of al-Teeb and Dwaireeg as estimated in the previous section will be considered. Since the flow is seasonal and controlled, it is concluded that 10% (or 10 (m^3/s)) will be available to replenish the marshes area. This assumption is based on the worst-case scenario [28]. Table (1) shows the possible flow from different sources and its distribution among the marshes and Shatt al-Arab. The total of the allocated flow is 470 m^3/sec. The total area of reflooded marshes is 4440 km^2. The flooded area is about 55.5% of the marshes area of 1973, and it's higher than any recovered area since 2003 but less than the planned recovery by the IMoWR. The water allocated to Shatt al-Arab is 128 (m^3/s), more than 2.5 times what is allocated to the river currently.

Table 1. Available water distributed among the three marshlands and Shatt al-Arab.

source	Flowrate (m^3/s)	Marshes (m^3/s)			Shatt al-Arab (m^3/s)	notes
		Al-Hammar	Central	Al-Hawizeh		
The Euphrates	178	100	78	0	0	e-flows
The Tigris	185	0	100*	35	50	e-flows
Al-Karkheh	19	0	0	19	0	e-flows
The Karun	78	0	0	0	78	e-flows
Al-Teeb and Dwaireeg	10**	0	0	10	0	Uncontrolled border flow
Total	470	100	178	64	128	
Total marshes recovered (km^2)	-------- --	1300	2310	830		

* This flow amount is delivered to the central marshes via al-Gharraf, Albtearah, and al-Uraieth equally.

** This amount may increase dramatically depending on the amount of rain falling on the related watersheds.

The water management plan, including the recovered marshland and the water distribution network along flowrates, is represented schematically in Figure (4). An important component of the plan is to isolate Shatt al-Arab from the marshes hydraulicly to prevent mixing waters from the two bodies. Separation is required to prevent marshes' drainage water from entering Shatt al-Arab. In order to achieve isolation, a hydraulic connection is established between the Hawizeh Marsh and the central marshes through an inverted siphon that spans the Tigris River in the Kasara Regulator area. The two marshes, then, connected to al-Hammar via an inverted siphon cross the Euphrates by the Chibayish town. The three marshes drain to the Gulf through Shatt al-Basra.

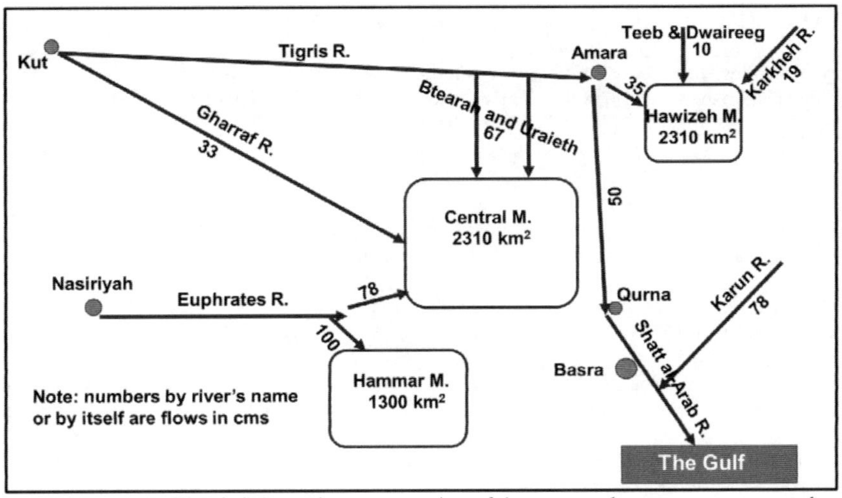

Fig. 4. Schematic representation of the proposed water management plan

5. Discussion

The water management plan hinges on the availability of the e-flows. E-flows implementation requires a basin-wide management approach. Such an approach is not practiced and may not be feasible for all river basins discussed so far due to hydro-politics dominating the area. Turkey, Syria, and Iraq share the Euphrates, while Turkey, Iraq, and Iran share the Tigris. Iran and Iraq, on the other hand, share the Karkheh River, as well as al-Teeb and Dwaireeg. The Karun is Iran's only river that flows across Iraq's Shatt al-Arab.. Robust collaboration and understanding among the riparian countries are required to allocate e- flows for the discussed rivers and to implement the proposed plan. An encouraging sign is that Turkey, which controlled the origin of the Tigris and the Euphrates, is including e-flows in its water resources management practice. Therefore, an integrated water resources management plan for each river basin that is acceptable for the stakeholder may be achievable and can be considered as the foundation of e-flows allocations.

Summary and conclusion

A water management plan that allocates constant water flow to restore the marshes and Shatt al-Arab is outlined. The two components of the proposed flow are: first, the environmental flow of the Euphrates, the Tigris, the Karkheh, and the Karun, and the second flow source is the uncontrolled valleys and sheet flow along the Iraq-Iran borderline. The marshes of southern Iraq, namely: al-Hawizeh, the Central, and al-Hammar marshes, are located within three governorates: Basra, Nasiriyah, and Amara. The total area is estimated at between 8000 and 20000 km^2.

Most of the marshes area has completely dried and transformed into bare land and salt crusts because of the previous regime desiccation works. Attempts to restore the marshes started in 2003 but with no meaningful success. Every 1000 km^2 of the marshland requires 77 m^3/sec to be restored and stay flooded year around.

The proposed plan allocated a flow of 470 m^3/sec, enough to restore 4440 km^2 of the previous marshland and Shatt al-Arab. The flooded area is about 55.5% of the marshes area of 1973, and it is higher than any recovered area since 2003. The water allocated to Shatt al-Arab is 128 (m^3/s), more than 2.5 times what is allocated to the river currently. The plan hinges on achieving basin-wide integrated management schemes and on a robust collaboration among the riparian countries.

References

[1] Hillel D 1994 Rivers of Eden: the struggle foe water and the quest for peace in the Midde East. Oxford University Press, New York

[2] Altinbilek D 2004 Development and management of the Euphrates–Tigris basin. *Water Resour Dev* **20**(1):15–33. doi:10.1080/ 07900620310001635584

[3] Rahi K A and Halihan T, 2010 Changes in the salinity of the Euphrates River system in Iraq. *Reg Environ Chang* **10**(1):27–35. https://doi.org/10.1007/s10113-009-0083-y
[4] Kliot N 1994 *Water Resources and Conflict in the Middle East. London*: Routledge
[5] Rahi K A and Halihan T 2018 Salinity Evolution of the Tigris River, *Reg Environ Change* https://doi.org/10.1007/s10113-018-1344-4
[6] Rahi K A 2018 Salinity management in the Shatt Al-Arab river. *Int. J. Eng. Technol.* **7,** pp128–133.
[7] UNESCO 2011 Report of the United Nations Integrated water task force for Iraq, *Managing change in the Marshlands, Iraq's Critical Challenge*
[8] UN-ESCWA and BGR (United Nations Economic and Social Commission for Western Asia; Bundesanstalt für Geowissenschaften und Rohstoffe) 2013 *Inventory of Shared Water Resources in Western Asia. Beirut.*
[9] Partow H 2001 The *Mesopotamian Marshlands: demise of an ecosystem early warning and assessment technical report*, UNEP/DEWA/TR.01–3 Rev. 1 Division of Early Warning and Assessment. United Nations Environment Programme, Nairobi
[10] Kolars J 1994 Problems of international river management: the case of the Euphrates. In: Biswas AK (ed) *International waters of the Middle East*—from Euphrates–Tigris to Nile. Oxford University Press, London, pp 44–94
[11] Buringh P 1960 *Soils and Soils Conditions in Iraq*, Directorate of General Agricultural Research and Projects, The Ministry of Agriculture, Baghdad, Iraq
[12] Harza Engineering Company 1963 *Hydrologic survey of Iraq*: final report, Baghdad, Iraq
[13] CEB (Consultant Engineering Bureau) 2011 *Shatt Al-Arab, final report* (in Arabic), College of Engineering, University of Baghdad, 101, (Unpublished).
[14] Ali Abdulkhabeer Ali Al-yasiri, 2011Assessment of water quality in Al-Chabaish and Al-Islah areas in Thi Qar governorat Univesity of Thi-Qar Journale, 7 (1), pp1-11.
[15] Ali Abdulkhabeer Ali Al-yasiri,2007 Study of the chemical and physical properties of different water sites in Suq Al-Shuyoukh and Al-Jbayish areas in Thi Qar Governorate, Univesity of Thi-Qar Journale.3(1),pp 1-7.
[16] Yasir R A, Rahi K A and Abudi, Z. N. 2018 Water budget for abu zirig marsh in Southern Iraq. *Journal of Engineering and Sustainable Development*, **22**(1), 25–33.
[17] Polservice 1980 *Shatt Al-Arab Project-Feasibility Report*, VII, Ministry of Irrigation, Iraq
[18] Anantha L and Dandekar P 2012 *Towards Restoring Flows into the Earth's Arteries*: A primer on Environmental Flows. Pune: River Research Centre, Kerala & South Asia Network on Dams Rivers and People. P 52
[19]The Brisbane Declaration 2007 Environmental flows are essential for freshwater ecosystem health and human well-being, in *10th International River Symposium and International Environmental* Flows Conference (Brisbane, QLD). Available online at: https://www.conservationgateway.org/
[20] Bussettini M and Vezza P 2019 *Guidance on Environmental Flows*—Integrating E-flow Science with Fluvial Geomorphology to Maintain Ecosystem Services; WMO-No. 1235;World Meteorological Organization (WMO): Geneva, Switzerland, pp. 1–52.
[21] TLWP (Texas living waters project) 2021 https://texaslivingwaters.org/environmental-flows/ (accessed on July 6th, 2021)
[22] IUCN (International Union for Conservation of Nature) 2021 IUCN (International Union for Conservation of Nature), https://www.iucn.org/downloads/water_briefing_eflows.pdf (accessed on July 6th, 2021).
[23] UNEP-Nairobi Convention/WIOMSA 2020 *Guidelines for the Assessment of Environmental Flows in the Western Indian Ocean Region*. UNEP, Nairobi, 79 pp.
[24] Smakhtin V and Anputhas M 2006 *An assessment of environmental flow requirements of Indian river basins*. Colombo, Sri Lanka: International Water Management Institute. 42p.(IWMI Research Report 107).
[25] Mirza MMQ 1998 Diversion of the Ganges water at Farakka and its effect on salinity in Bangladesh. Environ Manage **22**(5):711– 722. doi:10.1007/s002679900141
[26] Rahi K A. Al-Madhhachi A T and Al-Hussaini S N 2019 Assessment of surface water resources of eastern Iraq. *Hydrology*, **6**, 57.
[27] Scheumann W 1993 New irrigation scheme in Southeast Anatolia and in Northern Syria: more competition and conflict over the Euphrates. *Q J Int Agric* **32**(3):240–259.
[28] Al-Madhhachi A S T, Rahi K and Leabi, W.K. 2020 Hydrological Impact of Ilisu Dam on Mosul Dam; the River Tigris. *Geosciences*, **10**, 120.

Analysis of the environmental reality of the marshes and its sustainable development

Israa Mahmooed Mohammed Irzoqy(1), Lamia Flaieh Ibrahim(2), Hodoud Mohamed Abboud Al-Tufaily(2)

(1) Baghdad' Directorate of Education, Ministry of Education, Iraq,
(2) Department of Geography, College of Education for Human Sciences, University of Babylon, Ministry of Higher education and Scientific Research, Iraq. *Corresponding Author:* Znbg50544@yahoo.com (1)

Abstract

The study of the reality and concept of the environment of the marshes, both natural and human, and both sides complement each other in terms of the mutual influence between man and the environment, and that this influence refers to the process of adaptation practiced by man to the natural environment, which tries to harness for him to achieve goals, most of which are economic, which are not without harm to the environmental reality of the marshes Such as overfishing and destruction operations such as the drying of large areas, and the reason lies in the lack of interest on the part of governments and ministries concerned with protecting the marshes from various environmental violations and their attempt to develop them Its geographical location and its origin, as for the second topic, we touched on the most important environmental problems that the marshes suffer from in Iraq, and the methods of dealing with the deterioration occurring in the environment of the marshes were included in the third topic. In addition to the suggestions and conclusions and a list of sources and references.

Keywords: Marshes, destruction, environmental pollution, Poaching, Scientific Planning.

1. Introduction

The first text in which the marshes were mentioned in the Epic of Gilgamesh, written five thousand years ago in the Sumerian language. Gilgamesh is the king of Uruk of Warka. In the epic confirmed by biblical texts, we find Gilgamesh building a boat from papyrus reeds and coating it with tar. The marshes are inhabited by about one million Iraqi citizens belonging to the original Arab tribes. As for the inhabitants of the depths of the marshes, they are called the Madan, and their life is similar to the life of the nomads. The inhabitants of the marshes generally depend on fishing, rice cultivation and buffalo breeding[1].

The Iraqi marshes represent the largest wetland ecosystem of its kind in the Middle East and Western Asia. It consists of a group of connected lakes and swampy lands, in the lower part of the Euphrates and Tigris basin, which historically extended over an area of more than 20,000 km2 from Iraq and Iran [1].

One of the most prominent challenges facing the marshes is due to water scarcity, drought and the absence of water-sharing agreements with neighbouring countries, the deteriorating water quality in the Tigris and Euphrates rivers as a result of sewage and industrial pollutants, saltwater intrusion from the Arabian Gulf, the lack of effective policies from the central and local governments, and the lack of coordination between them to perpetuate, preserve, develop and advance the marshes, and they lack sustainable development for all life facilities in them. note that it possesses the elements of development such as arable soil, water resources, population, and others

1.1 Research problem

The research problem includes:

1.1.1 Does the marshes have a role in the environmental reality and is it possible to develop their environment sustainably according to their natural.

1.1.2 The drying of marshes and their role in damaging the environment and depleting its wealth, as the water share that feeds the marshes is reduced and the concentration of salts and chemical elements increase.

IOP Conf. Series: Earth and Environmental Science 1002 (2022) 012010 doi:10.1088/1755-1315/1002/1/012010

1.2 Research Hypothesis:

Human and environmental activities interact, and this, in turn, led to the study taking the environment a great deal. The marshes environment in southern Iraq is an integrated environment in terms of population, animal and plant resources and nature, creating an ecological balance in it. The marshes have a positive impact on the environment, as they are a good source of animal food and vegetarianism and has a role in its development.

1.3 Marsh definition:

Marshes are freshwater bodies covering vast areas in Iraq that have known a distinct human civilization dating back about 6000 years, in which diverse living organisms live, such as birds, about 81 species, along with fish, wild animals and livestock. It has forests and aquatic plants due to the flow of the waters of the Tigris and Euphrates rivers to it. The lagoon is a shallow water body less deep than the lake. In terms of area, some marshes have a surface area that is either larger or smaller than the lakes. The lagoon is characterized by large seasonal changes in area and depth [2]. The marshes are a group of wetlands or picturesque bodies of water fed mainly by the Tigris and Euphrates and extend in southern Iraq in a triangular shape between the cities of Al-Amarah in Maysan Governorate, Nasiriyah in Thi-Qar Governorate and Basra in the same governorate.
They are a group of water bodies that cover the lowlands located in the south of the Iraqi alluvial plain, and it is in the form of a triangle. The major cities within the marshlands are; Al-Amarah, Nasiriyah, Basra, Thi-Qar and Maysan. The area of land covered with water expands at the time of the flood, in late winter and during the spring, and the days of the Chihod shrink.Its area ranges from 35-40 thousand square kilometres. The early Arabs called these areas "Al-Bataeh", the plural of Buta'ha because the water in it fell, that is, it flowed and expanded in the ground and reeds were growing there. On July 17, 2016, UNESCO agreed to add the marshes to the World Heritage List as an international nature reserve, in addition to the ancient archaeological cities located near them, such as Ur, Eridu and Warka [3].

1.4 Causes of the deterioration of the marshes:

Despite all this environmental richness and ancient human history, the marshes - especially the central one - have been exposed for a long time, specifically since the mid-seventies, to various forms of neglect and dissipation, to the extent that these marshes are now threatened with demise or gradual disappearance. The reasons for this decline vary between historical reasons and external and internal reasons. The historical reasons are represented in the practices of the regimes that previously ruled Iraq, which is equal to what happened during the era of the previous regime when large parts were drained to stop the rebellion movement that spread in the region in the early nineties, with the ambitions and bourgeoisie of the American ruler of Iraq after the occupation, who harnessed all his efforts and interests To control Iraq's capabilities and facilitate the task of oil companies in obtaining concessions in the marshes and other Iraqi regions [4]. As for the external reasons, they are represented in the construction of several dams in Turkey and Syria. Consequently, the incoming water flows through the tributaries of rivers and waterways, which represent a lifeline for them, have stopped, as well as the lack of rainfall and water scarcity due to increased drought.
Its houses: They are built on the surface of the water completely from reeds, the area of which does not exceed the area of the house (the hut) that is built above it. Those islands are called (Ashman), which is a Sumerian word meaning hill. Some of them are made of mud-built on the edges of the marshes.

1.5 Disagreement over its origins and consensus on its importance:

The subject of the marshes has gained great importance, especially after its nomination to join the World Heritage List, which is the Al-Hawizeh Marsh in Maysan Governorate, as well as the West Hammar Marsh in Thi-Qar Governorate.
Marsh is a local term used to refer to swamps and marshes scattered near the mouth of the Tigris and Euphrates in southern Iraq. Underground water, which is usually close to the surface of the earth, the groundwater seeps into it and in this case, it is called swamps (Marches), but if the level of the ground is slightly higher than the level of the groundwater without interrupting it, then Sobkhas appear, which are wetlands that are difficult to cross. The soil of the marshes is fertile consisting of silt, heavy clay of water and wet bodies. Examples of these ground features are the Hammar swamp, which is about 28 km long and about 5 km wide. It is located near the confluence of the Tigris and Euphrates rivers in the southernmost part of Iraq. It is a huge swamp that extends in an east-west direction [5].

1.6 How the marshes arose:

Archaeological investigations confirm when the marshes were present and healthy and after drying them by the previous regime, that large parts of the marshes of the governorates of Nasiriyah and Maysan were dry land exploited for agriculture, and there were some permanent swamps confined, especially in Maysan governorate, and the evidence for this is the spread of archaeological sites in them. In the Hammar Marsh, whose area is estimated at one million and forty-eight thousand acres, we were able to install 122 archaeological sites between small and large, and the history of some sites dates back to the era of the dawn of the Sumerian dynasties, i.e. around 2800 - 2350 BC., Al-Sidinawiya, Kerma Bani Said, and these sites were either submerged in a marsh surrounded by water, depending on their height. As for the governorate of Maysan, there are 48 exposed archaeological sites, most of which date back to the Parthian and Sassanid eras, and some of them are older, dating back to the first millennium BC. These sites are spread in the Al-Hawiza marsh. Al-Wadiah, Al-Shaheen, Buraydah, etc., which are like its sisters in the marshes of Nasiriyah. Either it was immersed in the lagoon or surrounded by water. The views of the scientific community have varied, as one of them is likely to say that the sedimentary plain in southern and central Iraq was a sea in the Old Stone Age, i.e. about half a million years ago, and this sea began to retreat towards the south due to sediments and silt coming from the Tigris and Euphrates Rivers and their tributaries, and thus the plain appeared The sedimentary in its current condition, and the owners of this opinion used to say that the marshes in southern Iraq are the remains of that sea [6].

That the areas of the marshes were dry land on which the Iraqis lived and practised their economic and social activities and built their future civilization as in the rest of Iraq, In the same way, the two rivers and their tributaries were engulfed by a massive and unprecedented tyranny, and the major dams and irrigation projects that were neglected due to the weakness of the Sasanian state in the late days of its rule were destroyed in Nahrawan and the old great dam as well as the old Diyala dam and other areas. To wide marshes extending like the sea, these marshes became known at that time as Bataeh. Arab historians and townspeople described them extensively and included them on their maps. The rate of advancement of the delta was at a rate of (1, 6) km every seventy km and every thirty years during ancient times [7].

1.7. Location and area of the search area:

The marshlands are located between latitudes 50° 30' and 50° 32° in the north, between the Iranian border in the east, and the edge of the plateau in the west. Among the most important marshes in Iraq:

The governorates of (Maysan - Basra – Dhi-Qar) occupy the area of these three governorates (48,042) km2 and constitute (8,5%) of the area of the three governorates distributed as the basis of their small administrative units, amounting to 20 districts within the three governorates in which the marshes region is located, which are governorates (Maysan - Dhi Qar - Basra), which are considered the centres of spatial and population balance [8]. The surveyed Iraqi marshes include: first; Marsh Al-Dalamj, one of the marshes south of the Euphrates, second; The marshes of Maysan, third; The marshes in Al-Chbayish.

1.8 Marshes are divided into two groups: The first group:

1.8.1 First: The group of marshes located east of the Tigris River, the most important of which is Al-Hawizeh. Its area inside Iraq is about 2,863 km^2.

1.8.2 Second: The marshes located to the west of the Tigris, the most important of which is the Hammar marsh, which has an area of about 2441 km^2.

The Euphrates marshes extending from Al-Khidr to Al-Kifl between the two branches of the Euphrates (Al-Hilla and Al-Hindiya). It consists of several small marshes. The total area of the marshes, its estimates ranged between 9,000 and 20,000 km2, and the inhabitants of the marshes live on small natural or manufactured islands in the marshes, and they use a type of boat called Mashhouf for their movement and travel. The environment is considered a good source to provide a lot of food from fish, birds and agricultural materials that depend on the abundance and sustainability of water such as rice and sugar cane. Table (1).

Table (1) water areas of the southern Iraqi marshes for the year 2017

Marsh's Name	Area/ Km^2
Middle Marsh	1800
Hammar Marsh	2433
Al-Sanaf Marsh	11
Al- Hawizeh Marsh	2608
Total	6852

Source: 1- Satellite visuals of the American satellite Landsat (1.2.5.7) for different years and with different sensors from the previous version. Landsat 8. The distance and mapping were calculated based on the GIS program.
2- Ministry of Water Resources, Marsh Recovery Center, Studies and Design Department, unpublished data for the year 2012.
3-Hussein Alewi Naser Al Ziadi, The marshes of southern Iraq land and population, Alfaihaa for print and publishing, 2019.

The idea of draining the marshes:

Draining the Hammar Marsh appeared in the report of William Wilcox in 1911 when he proposed to implement linking the end of the Euphrates River in the Al-Tar area with the exit of the Hammar Marsh in the city of Al-Jbayish and placing a plug for this stream on both sides to deprive the Hammar Marsh of its most important water sources. This proposal was implemented in the eighties of the last century in the name of the excavator The draining of the Hammar Marsh also appeared in the report of the Tebet and Partners Company in 1958 under the name of the irrigation and saltwater drainage project, and it was exposed to drying in the nineties of the last century, specifically after the 1991 uprising or what is known as the popular uprising, as punishment for the marsh residents who rose against the former regime. Map (1). Only 4% of its total area remained after draining 96% of it. However, the current government has started projects to develop the marshes, as the marshes are now considered one of the most beautiful tourist areas in Iraq [9].

1.9 The importance of the lands of the marshes and the elements of sustainable development in them:

Sustainable development: It is the process of developing the land, cities, communities and all human activities in order to meet the needs of the present without compromising the ability of future generations to meet their needs. Iraq faces the danger of environmental degradation, which must be overcome without abandoning the needs of economic development, as well as equality and social justice.

Map (1) The location of the marshes in Iraq

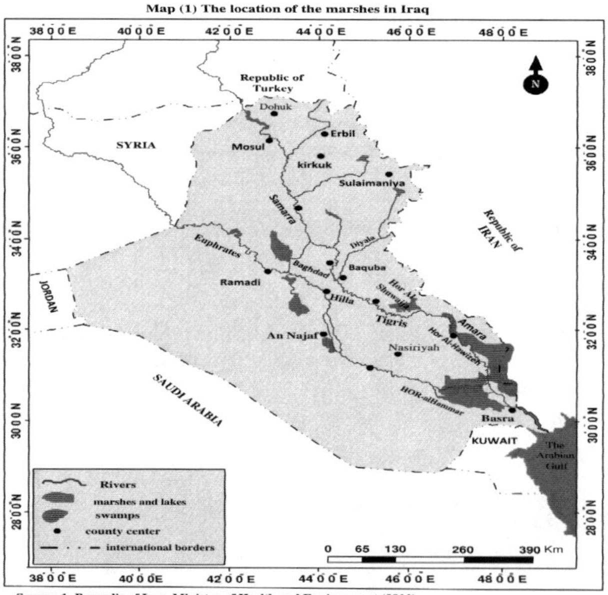

Source: 1- Repuplic of Iraq ,Ministry of Health and Environment (2011)
2- Ministry of Irrigation ,General Establihment for studies anddesign(G.E.S.D) ,Shatt Al – Arab project Feasibil Report Draft, sudies of salinity problems, part A .Text olservices co . Basrah, Iraq, 1979,pp151-155.

The importance of the marshes lies in Iraq, as it is a unique human community supported by natural resources for the unity of the ecosystem and possesses the elements of sustainable development.

1- The traditional way of life continued by the Arabs of the Marsh or Madan has remained this way for thousands of years.

2- The recorded history of the same region, as well as the beginnings of world history, beginning with the history of Sumer, which documented the reed houses (the hostel) and the longboats (the Al-Mashhouf) and it has been immortalized for more than 5000 thousand years.

3- The marshes provide an environment for the traditional agricultural production of millet, rice, and vegetables and the appropriate and permanent benefits of environmental resources, hunting birds and grazing cows.

4- The development of unique knowledge, which is how to manage and benefit from the natural environment of the water areas.

5- Animals and the diversity of living things:

The habitat of the marshes is of importance to a large number of living organisms as it has a worthy contribution to biodiversity and also contributes to supporting the migration of continental birds to the region because it supports the rare living organisms that are endemic to it in the essence of the important fact on the production of rare organisms and when they are lost they will be lost forever. The Hammar and Al-Hawizeh marshes together are likely to provide the environment for two-thirds of the wintering wild birds such as geese and ducks in the Middle East. They are among the important areas for the gathering of migratory waterfowl between the breeding grounds of Western Siberia and Central Asia, as well as the winter quarter in Africa. The marshes are one of the eleven non-marine water areas in the international classification as an endemic bird area, as they provide the world sector for these creatures with two types of creatures, the Basrah Reed Warbler and the Iraqi Babbler. On this basis alone, the marshes are of global importance to birds in terms of biodiversity [10].

1.9.1 The natural environment in the marshes and its role in sustainable development:

The Tigris and Euphrates rivers are the main sources of water in the marshes, in addition to the flow of several main tributaries from Iran towards the Tigris and Hawiza. From the top of the upper Mesopotamian plain to the Gulf, the rivers turn slowly.

Biodiversity and the marshes that it possesses are essential for sustainable development:

1.9.2 The environment of the marshes is home to many organisms that can be adopted in the development of the marshes, as more than 95 species of fungi live in it, more than 260 species of phytoplankton, parasitic and algae, (51) types of aquatic plants, and about (89) species of zooplankton and (92 of the large invertebrates and more than (41) species of migratory river and marine fish, including more than (159) species of protected and migratory waterbirds and more than (18) species of protected plants [11].

1.10 Obstacles to sustainable development in the marshes:

Some areas of the marshes suffer from the presence of war remnants and unexploded bombs, especially the marshes east of Al-Hammar, adjacent to the city of Souq Al-Shuyoukh [12] The exposure of the marshlands to the phenomenon of desertification is a destruction of the Earth's atmospheric energy, and it can lead to the end of desertification being complete, and it is a manifestation of the widespread deterioration of environmental systems that leads to a diminishing of the Earth's vital energy represented in plant and animal production [13]

The marshes have a positive impact on the environment, as they mean a good source for providing food resources from fish, birds and agricultural materials that depend on the abundance and sustainability of water such as rice and sugar cane [13] and fish is one of the most important elements of livestock in the marshes area, and its abundance and types are livestock, as they are Provides income for many fishermen who inhabit the marshes [14].

Human and environmental activities interact, and this, in turn, led to the study of the environment taking a large space, as it has become a problem threatening the rivers as a result of an increase in life demands, which led to damage environment and the depletion of its resources. The marshlands environment in southern Iraq is an integrated environment in terms of population, animal and plant resources and nature, which creates a balance Environmental Now the issuance of projects to investigate the marshes and drought water level in terms of the lack of the water share that feeds the marshes and the increase in the concentration of salts and chemical elements [15].

1.11 The natural components of the marshes of southern Iraq:

1.11.1 Climate influences:

The marshes region is characterized by being a hot, dry region in the summer, which includes six months from May to October, and is characterized by a long hot dry summer and shorter rest of the other seasons, with less than 200 millimetres of rain annually. The climate in the marshes is subtropical and dry. It has hot summers (average maximum temperature of about 43 degrees Celsius) and mild to cold winters (average minimum temperatures of 4 degrees Celsius). In summer, dust storms are common and heatwaves can raise temperatures to 48 degrees Celsius. During the winter, temperatures can drop to -8°C. Humidity is low while evaporation from water surfaces is high, which causes water loss because evaporation is ten times more than the contribution of precipitation in the region. It is noted that the highest temperatures were in July, followed by August, while June, in general, was the lowest of the three months in temperature [10].

1.11.2 Terrain:

Iraq consists chiefly of a flat, low-lying plain that rises slowly to the southwestern plateau, which is about 1,000 meters high, and the northeastern Zagros mountains, which reach a height of 3000 metres. Elevations on the nearly shapeless Mesopotamian plain are characteristic of most of the country and range from a few meters below sea level to about 400 meters above.

1.11.3 Water Resources:

The core of the marshes is centred around the confluence of the Tigris and Euphrates rivers. It is usually divided into the three main areas: the Hammar marshes, south of the Euphrates River; the central marshes (Qurna) between the Tigris and Euphrates rivers; And Hawizeh marshes to the east of the Tigris River. The marshes formed a series of nearly interconnected marsh catchments and lakes, which flowed into each other. During periods of high floods, large areas of the desert were submerged underwater. As a result, some of the previously separate marsh units have fused, forming a larger wetland catchment. The wetland watershed itself was made up of a mosaic of permanent and seasonal marshes, shallow and deep lakes, and mudflats that were regularly flooded during periods of high-water levels [16]. Also downstream, the Tigris and Euphrates rivers meet to form the Shatt al-Arab River, which flows into the Persian Gulf and is thus affected by tides. Before the 1970s, there was a complex system of natural channels where the Tigris and Euphrates formed the inner delta at the source of the marshes, and the marshes formed a system So closely connected that the rivers merged and disappeared. This system changed in the last century with the construction of agricultural canals, drainage systems and hydraulic structures, which greatly affected the hydrotherapy of the entire system. Water from the Tigris feeds the Hawizeh and Central Marshes due to the insufficient carrying capacity of its channel. The main Tigris canal also reduces its cross-section as it approaches the marshes; Where the flow of the river drops significantly. The Euphrates is restricted by higher dams from Nasiriyah to Qurna. However, there are secondary tributaries that connect the Central Marsh with the Euphrates via several canyons along the northern dam of the Euphrates. The Euphrates River divides into several arms downstream to Nasiriyah, and flows into the Hammar Marsh along several secondary channels; The main tributary of the Euphrates flows towards the confluence of the Tigris with the Qurna, but it is currently closed by a barrier that facilitates its flooding through the wetlands. Water flows from the Hammar Marshes to the Shatt al-Arab and the Shatt al-Arab and Shatt al-Basra rivers through Qarmat Ali, while the central marshes flow up the Euphrates and then, in Qurna, to the Shatt al-Arab. The water flows from the Hawizeh Marsh to the Tigris and Shatt al-Arab [9].

1.12. Iraqi river

1.12.1 Euphrates River:

It is the largest river in Mesopotamia, with a length of about 2,800 km, and it begins with a thousand km in Turkey, where it originates in the mountainous region and descends to cross the Syrian Republic and continues to the southeast inside Iraqi territory. Its length in Syria is about 675 km, and in Iraq, it is 1,200 km. The rain and snow falling on the upper edges of the river basin are the main sources of its water[10]

1.12.2 Tigris River:

It originates from the Anatolian plateau in southeastern Turkey, to the eastern Taurus Mountains and the Zagros Mountains in Iran, and ends in Iraq after passing a short distance in Syria. There are several dams on the Tigris River, larger than the Euphrates River.

1.12.3 Shatt Al Arab:

As for the Shatt al-Arab, it is a waterway that extends about 120 km, starting from Qurna and pouring its fresh water into the northern Gulf. Qurna is the area where the Tigris and Euphrates rivers meet, which stem from the mountains of Turkey and cross in parallel the Turkish and Iraqi lands. At the confluence of the two rivers, the marshes were found. It is the result of natural processes over millions of years due to the exposure of certain areas in the triangle of Nasiriyah, Basra, Al-Amarah, to the flooding of the two rivers, and this water combines, creating a distinctive environment [9]. It is the main nutrient for the waters of the northern Arabian Gulf, and its waters are characterized by low salinity and an abundance of dissolved nutrients such as nitrates and phosphates [6].

1.13. Third river:

Iraq completed the Third River Project in early 1993, diverting the course of the Euphrates River to prevent it from feeding the Hammar Marsh with fresh water. Its length is about 565 km and its width is 90 m. The third river carries the wastes of agricultural areas in central Iraq and the floodwaters to the northern Arabian Gulf region through Khor Al Zubair and then Khor Al Sabiya. If the legends that the long-lived residents of the marshes are still circulating, which go by saying that some Sassanid kings during the fifth century AD wanted to drain the marshes to be vast fields for the cultivation of grains, and they could not achieve that dream.. the former Iraqi regime did not dream of drying the marshes Except to eliminate the focus of the Iraqi resistance there. The marshes of southern Iraq almost disappeared from the map of the country, due to the crime committed by the architects of the former Iraqi regime against the oldest environmental systems in the world. This crime, which was carried out with great coldness and indifference, was taking place continuously despite the air embargo imposed by the coalition forces to protect the southern region located on line 32.

1.14 Human Ingredients (Population):

The inhabitants of the marshes are among the first in Iraq, and the rate of decline began to increase due to migration to neighbouring cities, especially to the centers of the southern governorates such as Basra and Nasiriyah. They have names such as (the inhabitants of the marshes - the Arabs of the marshes - the Madan), and most of the inhabitants live in floating banks that are separated from others by a waterway where they communicate with each other by means of (the Mashahif), which is similar to a boat used for transportation and work.
This style of living is unique and similar to the settlement pattern of some villages in Southeast Asia on the banks of rivers and coasts. This style can be developed with the available ingredients to be able to promote it and overcome the living difficulties experienced by the population. The United Nations report described the Marsh Arabs as a distinguished and ancient people and that their lifestyle, which dates back to five thousand years, has become threatened with sudden extinction, this lifestyle that is the legitimate heir of the Sumerian and Babylonian civilizations. "The destruction of the marshes has devastating effects on wildlife, and this would have negative impacts on animal diversity at the local and regional levels.

1.15 The economic and environmental importance of the Iraqi Marshlands:

The marshes are of great importance from the traditional agricultural point of view for the people of the region as they are used for fishing, hunting birds and grazing livestock. It is considered a nursery and hatching area for some creatures and birds that are of commercial and environmental importance. The environment of the region filters natural and inorganic pollutants from the water, and accordingly, the water that flows into the northern region of the Arabian Gulf becomes purer than its counterparts in that environment.
The preliminary economic estimation of the value of sedge and traditional agricultural activities alone in the marshes is estimated at more than 300 million US dollars. The annual rate of fishing from the marshes is about 1700 tons of small fish. Shrimp and agricultural nurseries in the marshes may constitute more than 40% of the total area of Kuwait. The universality of animals and the diversity of different neighbourhoods are of fundamental importance to the preservation and preservation of unique cultural heritage, of importance to the preservation of the quality of the environment.

15.1 Fishing places:

The marshes are an important habitat for diversity and diverse species in two respects, economic and scientific interest. Some species migrate for economic importance outside the borders of Iraq, especially the small fish called in Latin Metapennaeus affinis, which is of great economic importance to professional fishermen on the northern Gulf coast. A large number of marsh organisms of the cyprid family are of great importance to the study of evolutionary theory. The marshes in southern Iraq have a long history that extends back to more than 5,000 years, and they are of great environmental importance, as they possess many of the elements of : -
1- They represent areas for tourism and it is possible to invest all the ingredients in them.
2- It forms a way for migratory birds to cross between the continents.

3- Sand warehouse for the continuity of the freshwater fish fishing areas, as they constitute ideal hatcheries and nurseries for fresh and even marine fish species such as pilgrims and shrimp.
4- It supports the marine ecosystem and marine life in the Arabian Gulf.
5- It protects endangered animal species by providing them with nutrition, shelter and reproduction.
6- The marshlands are natural sources of water across the border.
7- It constitutes an unparalleled human heritage and has been home to the indigenous people for 5000 years (Sumerians. Ur. Eridu. Lagash.)
8- Marshes contribute to the disposal of water pollutants through their deposition and representation of the natural life cycles of nutrients, plant and animal organisms in the food web.
9- The presence of its oil reserves and military conflicts have led it to face major challenges in maintaining its ecosystem and exploiting its resources for growth and development. [17]

1.16 The marshes of southern Iraq and their importance:

Marsh al-Hawizeh, which overlaps with the Iranian border, the Hammar marshes, which is in the middle of the region, and the Euphrates marshes, which extend and expand towards the north and west of Basra and the south of the al-Amarah region, reaching close to the mouth of the Tigris and Euphrates to mix and form the Shatt al-Arab. The area occupied by the marshes of southern Iraq can be estimated between 15,000 square kilometres. - 50,000 square kilometres because the water area of these marshes is not fixed and unstable. It expands and contracts according to the rhythms of the seasons. In the autumn, when the water levels of the Tigris and Euphrates are at their lowest, we find that the area of the marshes shrinks significantly, to return and expand in the spring with the season of melting snow in Both Turkey and Iran.The importance of the marshes includes rare animals and birds that cannot be found in any other part of the world, in addition to its wealth of fish, oil and other virgin minerals that have not been exploited yet. It also constitutes a kind of ideal nature reserves that receives huge numbers of migratory birds due to the abundance of the necessary food sources. for such birds.

The geographic and natural characteristics of the permanent marshes, represented by the geographical location, area, extension, climatic characteristics, the vital environment, and the diverse and diverse plant groups it contains, as well as the diverse animal groups in them.

The efforts made to include the marshes of southern Iraq on the World Heritage List opened new and promising horizons to restore life to these beautiful marshes in all fields. This requires drawing up a comprehensive policy for developing the area and completing the infrastructure to benefit from this decision in restoring and developing the marshes and ensuring their proper sustainability.

The Iraqi Marshes represent the largest wetland ecosystem of its kind in the Middle East and Western Asia. It consists of a group of connected lakes and swampy lands, in the lower part of the Euphrates and Tigris basin, which historically extended over an area of more than 20,000 km2 from Iraq and Iran. [18] included many types of plants Table (2)

Table (2) Names of plants in the marshes of southern Iraq

The scientific name	**Potamogeton pectunatus**
Phragmites australis	Vallisneria spiralis
Typha australis	Potamogeton lusens
Schoeno plectus	Potamogeton perfoliatus
Najas sp	Ceratophyllum demersum
Myriophyllum verticillatum	Hydrela
Salvinia natans	Lemnna gibba

Source; The Republic of Iraq, Ministry of Health and Environment, Dhi Qar Environment Directorate, Environmental Reality 2015. Hussein Alewi Naser Al Ziadi, The marshes of southern Iraq land and population, Alfaihaa for print and publishing, 2019.

1.16.1 The most important components of the marshes:

1. The environmental characteristics of the southern marshes indicate that it is a natural reserve due to the weak human influence therein.
2. The water landscape, plant and animal life, and unique human life make the marshes a tourist attraction[19].The environment of the marshes has contributed, since the early days of human life, to the making of the elements and components of civilization through which humanity moved to historical times, as confirmed by archaeological evidence. In the marshes, the first walled cities were built, such as Uruk, Eridu, Ur, and Lagash. In the marshes, writing was invented for the first time in the history of the world, marking the start of the written history of the world. From there, the systems of government, laws,

laws, arts, industry, irrigation, agriculture, religious beliefs, literature and music began. Through which he emphasized that the environment of the marshes represents a lifestyle that has its own characteristics that distinguish it from other forms of living and coexistence in Iraq ((eco-viability)) with its manifestations (distinctive, luxurious and transformed). A special style of life and culture in it, and according to the consensus of specialists, this environment, in the opinion of the researcher, combined heritage, thought and civilization, which eventually reached high peaks.
The marshes, in their cultural and historical form, constitute the appropriate incubator for the reconstruction of the Iraqi personality, and that the provinces and universities in which the marshes are located have the responsibility to preserve them as a natural reserve whose water produced man and his civilization. Map (2)
Thus, it may turn into a global tourist resort frequented by the people of the world, in which the twenty-first-century person will ease his concerns and return to living with nature, in nature, and nature [1].

1.16.2 Obstacles to sustainable development in the marshes:

1- Lack of interest in the governmental situation at present and the present, and its shortening of the environment of the marshes with its natural sides, plans, plans, plans and governmental plans inside Iraq from the singularity of interest in the marshes
2- Construction of dams, and conditions for obtaining oil prices that are protected for the Marshes.
3- The effects of the Iraqi-Iranian war and the resulting displacement and displacement of hundreds of families. This war created a huge displacement crisis in which the social structure of the marsh dwellers was often cracked and collapsed.
4- The environment of the marshes is polluted with thousands of tons of mines, rockets and unexploded bombs, many of which have remained until today a source of death or disability for the residents or their attachments of unconventional weapons.

Map (2) the location of Iraq between neighboring countries and the location of the marshes in Iraq

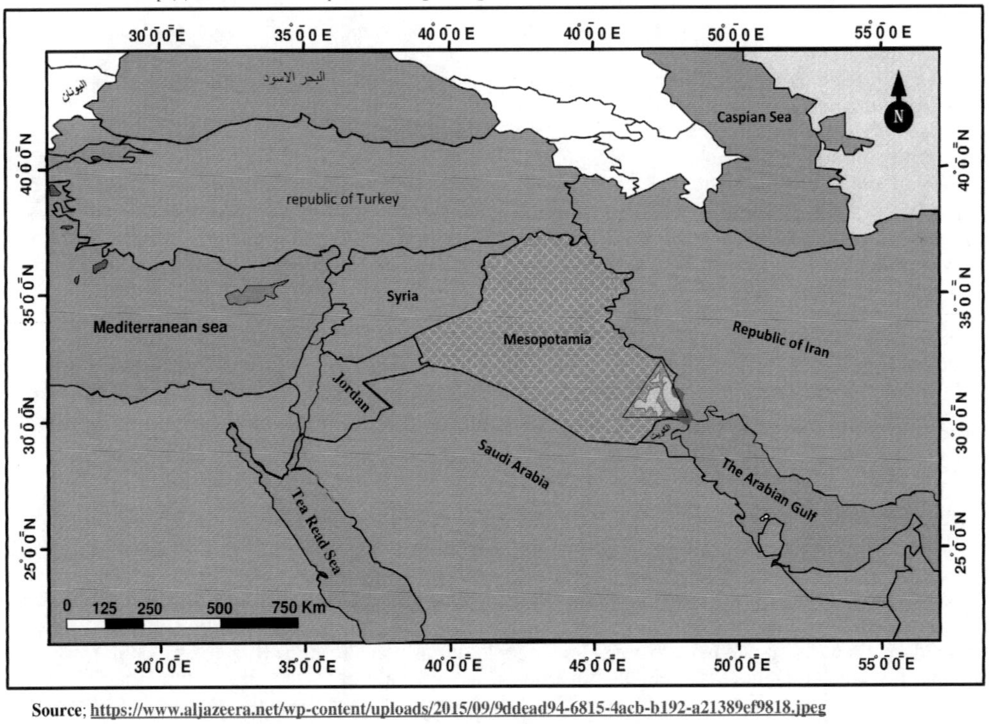

Source; https://www.aljazeera.net/wp-content/uploads/2015/09/9ddead94-6815-4acb-b192-a21389ef9818.jpeg

1.17 Factors that contributed to the increase in the problems of the environment of the marshes at present:

1. Classifying the marshes as mere (water and plants), based on which the task of developing the marshes was entrusted to the Ministry of Water Resources first and the Ministries of Environment, Works and Municipalities in the second degree.

2. The absence of central planning and the lack of coordination between the central government and local governments concerning the file of marsh management and development.
3. Weak state control and monitoring over the global grants that were distributed to official and non-official bodies, and which were often exhausted in marginal and meaningless programs and plans for the development of the marshes [20]
4- The social values of the marsh dwellers after the opposite migration that occurred after 2003, after which the most important variables in the process of developing and organizing the environment of the marshes and that these social values of the children differ significantly from the values of the parents as a result of several factors, including the factor of education and social interaction with communities and other groups outside the marsh areas, which Contributed to the acquisition of new trades and professions that were not known in the environment of the marshes previously.
5- The interaction of young people and their ideas with human values as a result of the communications revolution and the multiplicity of social media, which allowed individuals to communicate with different cultures in society[21].

1.17.1 Marsh Community Development:

The marsh community is a unique community with many characteristics that are not available in the corresponding areas supported by natural resources, which is the traditional way of life that has preserved its nature for thousands of years, starting from reed houses to wooden boats (Al-Mashhouf) and other methods of transportation, to ways of preparing food and ending with relationships The family and the nature of the local dialect, and this reality requires specialists, planners and decision-makers to advance the reality of these areas economically and socially and to draw a clear strategy and objectives that include giving guarantee and confidence to the population and social leaders in them not to prejudice the local character and social values to ensure the success of the plans developed through the participation of the population in preparing and implementing them and planning for it [22].
The development of the tourism sector and its development is very important to change the economic and social reality of the residents of the marshes for the better. The development of the tourism sector requires development programs, foremost of which are the education of the people of those areas on tourism behavior and the construction of infrastructure such as hotels, restaurants and cafes with building materials from the products of the marshes and the development of programs for hunting and fishing festivals Rural singing, plastic art and folklore, reducing negative influences, enhancing economic benefits, and developing tourism activity in the southern marshes, which is part of the regional map of economic and social development for the marsh dwellers in particular, and for Iraq in general.
This activity will contribute to providing wide job opportunities for the people of the Marshlands. In addition to the need to provide full support to the private sector to contribute effectively and to overcome legal and administrative obstacles in front of it, and on the prerequisite that there should be standards for work and an investment map with clear lines and open objectives. [23]Tens of thousands of people who used to live in the marshes area in southern Iraq live in poverty. Since the nineties, after the former Iraqi regime drained the water that was the mainstay of life in the region, whose civilizational history dates back to five thousand years[24].The marshes, which are the largest water bodies in the Middle East, were destroyed. The marshes, the largest of the water bodies in the Middle East, were destroyed, and the marshes slowly turned into a drought-stricken wasteland, and the marsh Arabs still remember the sight of their animals dying and their children sick with the disappearance of the water around them. But now the world's worst environmental crisis has begun to unravel as people, birds and livestock are starting to return.

1.17.2 The dangers of environmental disasters in Iraq:

Reports indicate that the water level of the Tigris and Euphrates rivers has declined to 43% compared to normal rates, a level that has been less than that since the beginning of the general water level recording. watering first, Secondly, the high level of salinity of river water, and the increase in the pollution levels of groundwater and running water due to its exposure to the flow of sewage and wastewater, and the high levels of pollution of river water that make access to clean water for human consumption difficult, especially with the decline of health services in general and the state of drought and the decline The rains led to the deterioration of the agricultural seasons. Iraqi agriculture, which already suffers from a deficit in meeting the needs of local demand, including grain crops.The deterioration of the condition of the natural pastures and the exposing of livestock and herders to more suffering due to the difficulty of providing fodder and the weak contribution of the concerned authorities to assisting those affected, which puts about 400,000 herders at a new risk, creating a new type of competition for natural resources and eliminating the new generation of livestock.
The marshes is the result of wars and as a result of the policies of the state's management of water in Iraq through the construction of dams and reservoirs at the meeting point of the Tigris and Euphrates, as well as the wrong practices carried out by many of the inhabitants of the marshes and adjacent area, including overfishing of fish and birds through the use of poisons Electrocution and these crude operations have led to damage to the vital system of the marshes.

1.17.3 Sewage disposal:

Pollution of the marshes is represented by the discharge of sewage water into the rivers that it irrigates up to the marshes. The marshes in southern Iraq have been exposed to various types of environmental climatic conditions that affected the ecosystem, especially after drying, where the environmental balance was disturbed and the character of the region changed, as well as changes in climatic and local conditions. More than (9555) km2 of marshes and lakes quickly and with a direct impact on the local climate, and the area has become a dry environment and its exposure to the phenomenon of desertification, which is a process of demolition and destruction of the vital energy of the earth and can lead to a wide deterioration of the environmental systems that lead to the erosion of the earth's air represented by plant and animal production [25].

1.17.4 Overfishing:

Hunting in a healthy environment, such as fishing with electric current, burning areas of reed and papyrus lands, and using grains in them, to catch the largest number of animals is the black water platform, pink flamingos, hamlets, and others.

1.17.5 Vegetation:

The aquatic plants within the marshes are a major part of the food chain of other living organisms, and the vegetation cover death to more than 75% of the vegetation cover. [26] The dried lands turn into a desert of hardened mud that no longer accepts animals or natural plants, and it is one of the worst environmental disasters in the world now, and that the effects of this will include the Gulf region because the marshes incubate several types of birds and some fish hatch in them, which will have far-reaching damages. On the ecosystem of the marshes, it was completely destroyed and the balance between water, crops, and animals was disturbed. Picture (1), which found thousands of years ago that the process of rehabilitating this area is very difficult, as the marshes that desertify no longer accept life.

Picture 1: Al-Mashhouf permeates the waters of the marshes as a means of transportation surrounded by plants Source; https://usercontent.one/wp/hdf-iq.org/wp-content/uploads/2017/09/Marsh_Arabs_in_a_mashoof-400x242.jpg

1.18 Drying the marshes:

The Iraqi plan, which began in late 1991, is summarized in the revival of the Third River Project, whose idea dates back to the mid-fifties of the last century by digging the old course of the Euphrates River and diverting the waters of some of the tributaries that feed the marshes with freshwater to this artificial river and proceeded to drill it at great depths To ensure its withdrawal of groundwater sources due to its location lower than the depth of the marsh to withdraw the water of the marsh by osmosis and to ensure the drying of these fresh lakes, which extend over a wide range of water, so that this river undertakes to draw the water of the marshes from its sources as well as diverting the streams of many of the tributaries that fed them to dry it and eliminate the thick reed and sedge forests, which became an ideal shelter for those outside its control and opponents of it. The project reached its climax at the end of 1992 of the project, and after about 90 per cent of the areas of those virgin areas dried up, and since the implementation of the most important part of these projects began after the uprising in the south in March 1991, the volume of water bodies and vegetation cover decreased to less than 20 percent. Sources estimate that the population of the region has decreased from 750,000 to less than 50,000, confirming that the number is constantly declining and the population has moved to live in poverty belts surrounding the cities of Basra and Baghdad after their natural sources of livelihood dried up and their livestock died, . Experts in the affairs of the marshes called for an international campaign to declare the marshes area a natural environmental reserve before it was destroyed by Iraqi procedures [27].

1.18.1 Turkish dams:

Until the 1970s, the freshwater marshes covered up to 20 square kilometres, and in addition to providing the Marsh Arabs with a suitable environment for agriculture and fishing, they fed up to 40% of Kuwaiti shrimp fisheries. Its chicks head downstream into the bay, but then dam and drain projects cut off the water supply.

His studies and calculations showed that the Turkish dams built in the late eighties and early nineties reduced the level of the Euphrates River by more than 20% and could eventually reduce it by 50%.That the water storage capacity of all dams on the Euphrates is six times greater than the annual flow of the river, and the dams that Turkish plans to build on the Tigris, including the controversial Ilisu Dam, which is behind British companies, will further reduce the flow of water into the marshes.

1.18.2 Environmental Effects of Drying the Iraqi Marshes:

Many harmful environmental effects result from draining the marshes, the most important of which are the following:

1- The decline in groundwater levels and the reduction of water bodies means the transformation of areas of wet cohesive lands into dry, loose land that can be eroded by wind, which means creating conditions for the spread of sand and dust storms and the movement of sand due to the prevailing northwest winds and not to mention the salts and vapours carried by the wind, which move with sand and dust.
2- Depletion of soil fertility due to the volatilization of fine grains of green and clay, as well as the organic matter originally present in the soil.
3- The death and destruction of plant clans and the birds and animals that live on them.
4- The disappearance of marine creatures and plankton.
5- An imbalance in the water balance.

Map (3): The main marshes of southern Iraq

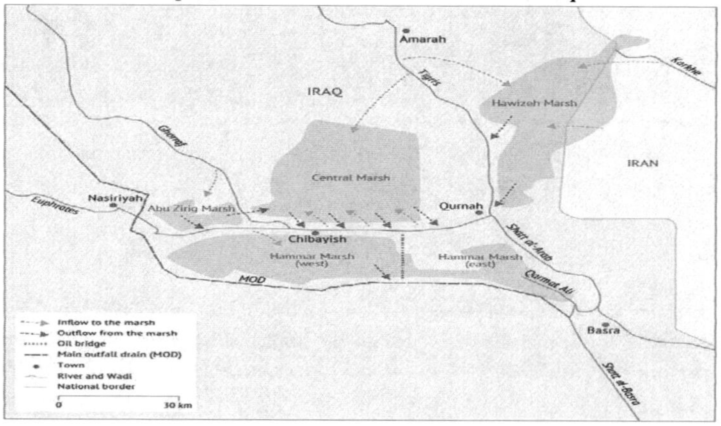

Source; https://water.fanack.com/wp-content/uploads/2017/05/Fanack_Irak_Map-02_705px.jpg

There is live evidence of the environmental destruction that the drying of wet areas brings, as the drying up of the areas surrounding the Aral Sea in Kazakhstan in the western part of the former Soviet Union exposed an area of 4800 square km to the winds, which turned the area into a huge theater for active wind operations where continuous dust storms and the transfer of salts to areas And that was during the period from 1961 to 1964. As for southern Iraq, the Iraqi regime claims that the draining of the marshes is due to the construction of dams in Syria and Turkey in the upper reaches of the Tigris and Euphrates rivers. Wet areas, consisting of swamps, marshes, and other ecosystems, are considered the most sensitive and the most vulnerable. They are, on the whole, a delicate system in which complex hydrological, biological, climatic, dynamic and environmental factors overlap. This prompted the US Environmental Agency to establish specialized departments responsible for supervising the management, control and regulation of the exploitation of these ecosystems.

That the mammals and fish that live in the area are now considered extinct, and the quantities of fish in the northern Gulf, which depend on the marshes for fish reproduction, have decreased sharply. There is also a type of large rat, known as the pig rat, which is unique to the region has become extinct. A unique way of life is on the way to extinction, but the United Nations has not completely given up on the situation, as confirmed. The situation in the Iraqi marshes seems bleak, but there are examples of similar areas in the world recovering after their disappearance. That this happened in Cameroon and the United States, And that there is an urgent need to preserve what remains of the marshlands on the Iraqi-Iranian border, and to reconsider the dams that were built to prevent floods in the fifties. These dams have become useless and if they are removed, they will make way for the marshes to be revived again. However, demolition is always easier than building, and marshes represent thousands of years of natural development. The regional effects of draining these swamps were devastating to the environment in the northern Arabian Gulf region, as it included an increase in pollution with hydrocarbon compounds, trace minerals and pesticides, and an increase in sedimentation in its marine environment, which resulted in direct negative effects on fisheries, coral reefs and human health. And adverse effects on regional wildlife, which resulted in a significant decrease in the numbers and species of endemic birds accompanied by a significant increase in the blooming of harmful algae and fish deaths in the region, in addition to the occurrence of regional climatic changes accompanied by an increase in temperature and rates of dust storms, which contributed to the rise in temperatures Earth and global warming [27].

1.19 The effect of draining the marshes on animal husbandry, the most important of which is the buffalo:

The buffalo is one of the most important animals in the marsh environment. Its upbringing occupies the main craft practiced by the ancient population. It is raised by a population called (the metal), and it depends on buffalo milk by making milk products and making use of its meat and offal as fuel. The Iraqi buffalo is distinguished by its adaptation to the environment in which it lives in the marshes. The Iraqi buffalo is one of the distinctive species in the world. The number of buffaloes in the world is 3 million, 90% of which are concentrated in Iraq. It is compatible with the environment of the marshes, and that the buffalo animal is threatened with extinction in Iraq in the absence of attention to the marshes and continue drying them Picture (2)

1.19.1 Marine environment (fish);

The marshes provide a suitable environment for the presence of water, sunlight and food resources on which plant organisms such as algae, crustaceans and insects feed, due to the absence of sea currents. For food the population and their commercial activity. The marine environment is affected by any changes that occur in the northern region of the Arabian Gulf, and the discharge of pollutants into the marine environment reduces the chances of development for different marine organisms and causes changes among its various elements that may cause damage to the human morsel through the food chain, and the matter gets worse the more it increases Focus on the long term. Therefore, we see that the drainage of the Shatt al-Arab waters constitutes a major and important source of nutrients, which include nitrates, nitrites, phosphorous compounds and silicon, and that any increase or decrease in the amount of nutrients affects the balance of the biochemical cycle of nitrogen and phosphorous in the marine environment, as nutrients are an important food source for algae and plant vermin in the marine environment. photosynthesis processes. The effect of drying the marshes in southern Iraq has a negative impact on the marine environment and its lack of development, as the results of studies indicated an increase in the concentration of petroleum pollutants associated with suspended and sedimentary particles and an increase in the rates of pollutant deposition in the marine environment.

Picture 2: The Iraqi buffalo animal in the middle of a carrot in the marshes of southern Iraq
Source; https://media.voltron.alhurra.com/Drupal/01live-106/styles/1984x1200/s3/2021-05/2015-02-12T000000Z_943294106_GM1EB2C16R901_RTRMADP_3_IRAQ-DAILYLIFE.JPG?itok=w_-RLEaL

Birds: The marshes constitute a suitable environment for the presence of birds, where the conditions of the climate are suitable, water and food. - and geese) and these types of birds can be made as a reserve, developed and preserved in the environment suitable for their living.

Effects of the spread of desertification in the marshes:

The phenomenon of desertification has spread to reduce the areas of the marshes and expose large parts of them to salinization and the drying of water on their lands This resulted from the problem of salinity in the soil, which is a manifestation of desertification and because of the high level of saline groundwater and its proximity to the soil surface And then the accumulation of salts on its surface after its evaporation, leaving salt deposits with an area of about (100), (168) and (376) km2 of the three marshes, respectively, which led to the salinization of the soil and its drying and holding of its particles, which is a manifestation of desertification.

1.19.2 Environmental Effects on Agriculture in the Marshes of Iraq:

The areas planted with crops have been reduced for extensive measures from the marsh to salinization and to change the water quality, as the areas planted with rice have been occupied due to the decline of the marsh's water from areas and its cultivation and the high salinity of the soil and its replacement by the cultivation of wheat and barley crops that can tolerate salinity, in addition to palm and vegetables and it has shrunk to (13%) due to the low hydrological characteristics and the difference in the sources of nourishing water. The salinity rate of the marshes is (14, 15) dc/m, which limited the possibility of using water for different purposes and the effect of saline sewage drained from the public estuary. Which led to the pollution of the waters of the marshes and its low efficiency in agricultural investment [28].

Picture 3: The manifestations of desertification in the marshes environment in southern Iraq
Source; **https://upload.wikimedia.org/wikipedia**

1.19.3 Hard facts:

The environment of the marshes has facts, namely that environmental problems know no borders, and their impact does not appear in the short term, and any defect in the physical, chemical and biological properties of the ecosystem affects the systems related to it, such as the northern marine environment of the State of Kuwait and the nutritional relationships it contains between organisms represented by the food web of the region As a whole, scientific studies conducted on sabour, zubaidi and shrimp fish in Kuwait proved that they depend on the environment of the marshes and the areas near the Shatt al-Arab. Most of these fish migrate from the maritime borders of Iraq to the northern waters of the Gulf. Therefore, the fish stock of Kuwait was affected by the drying of the marshes, and there are significant changes that occurred in the water, a sharp drop in salinity in Khor Al-Subiya and North Failaka from 36 parts per thousand in 1982 to one part per thousand in 1996 and 1998, and the concentration of the northern suspended sediments more than doubled, as it was observed that species near the mouth of the third river affect the marine food chain in the region, disturbing the ecosystem in addition to the newly formed environmental sediments. Consequently, the water affected the Kuwaiti fish wealth, the most important of which are:

First: Establishing a new course for the Euphrates River and diverting it from the Hammar Marsh. This river carries the remnants of agricultural areas, such as fertilizers and pesticides, from the central Iraq region as well as the flood area to the northern Arabian Gulf region through Khor Al-Zubair and then Khor Al-Subiya.

Second: The construction of 22 dams at the Tigris and Euphrates rivers, called the Turkish Anatolia Project, with the aim of generating electric power and providing irrigation water to Turkey. In 2010, the rate of fresh water that flows into the Arabian Gulf will decrease. Followers of this issue believe that the situation in the Marshes of Iraq will have many side effects on the Iraqi and Kuwaiti environments, but rather on the environment of the region as a whole Therefore, observers see the need for the experts of the two countries to work in coordination with international organizations to develop studies to evaluate the situation, conclude agreements that prevent construction works harmful to the environment, coordinate at the international level, pump water into the Iraqi marshes, and abide by international environmental laws and legislation [29].

Picture 4: Types of marsh plants and their growth methods
Source: Hussein Aliwi Nasser Al-Ziyadi, Some Development Indicators in the Marshes of Southern Iraq, College of Education, Wasit University, Special Issue for Research of the Fourth Scientific Conference 2011, p. 949.

1.20 **Restoration actions**: several restoration actions have been implemented by both governmental and non-governmental institutions which include:

1- Marshlands restoration activities included surveys, site visits, modelling, water characteristics study, and infrastructure needs assessment.
2- Providing a technical evaluation in June 2004 in cooperation with the Ministry of Water Resources and the relevant university.
3- Implementation of a strategy for the restoration of the marshes that includes monitoring the ecosystem, the reconstruction of water bodies, the making of models to study the water and hydraulic characteristics, agricultural work on a large scale, fishing and aquaculture. Under suitable conditions, livestock breeding, dairy production and primary health care.
4- Developing capabilities through laboratory development, study tours, training, and forming a strategy for international aid to donor countries and participation.

Suggestions and recommendations:

1- Commitment not to build and construct inside the marshes any buildings with materials of bricks, plaster, cement, iron and wood, and to build the necessary buildings near the marshes of reeds, papyrus and Golan only to continue to preserve them as a nature reserve.
2- habits in society by enacting an appropriate law for that.

3- Protecting the environment of the marshes from pollution and the use of overfishing methods using poisons and electric shocks to birds and fish.
4- Protecting the tourists as well as working as a tourist guide at the same time.
5- The Iraqi universities concerned with the environment of the marshes, especially Basra, Maysan and Dhi Qar, should establish a special department for the environment of the marshes, aimed at establishing knowledge, skills and information that preserve and maintain this environment.
6- Encouraging governmental institutions and bodies and civil society organizations in the sectors of education, health, environment and human rights to hold courses and field workshops to rehabilitate the population in these areas and to achieve environmental sanitation represented by preventing the danger of pollution and organizing cultural awareness campaigns among the people of the marshes whose mission is to establish friendship with the environment The marshes and emphasis on the correct behaviour of fishermen and birds, as well as the correct methods of grazing buffalo, cows and other animals
7- Encouraging the scientific research process in the marshes, especially the social, anthropological, demographic and economic studies, which the marshlands environment in Iraq lacks.
8- Removing the marshes administration file from all forms of partisan and clan quotas and adopting the criterion of efficiency and integrity as a basis for its management.
9- Benefiting from the decision of UNESCO and the Ramsar Treaty to put pressure on Turkey, Syria and Iran to conclude a water-sharing agreement and with it the marshes' share.
10- Encouraging tourism investment near archaeological sites near the marshes, and enacting a law that facilitates investment and overcomes the current obstacles, provided that it takes into account the standards of world heritage and the establishment of tourism infrastructure (such as artificial islands within the marshes with tourist cities that include hotels, restaurants, casinos and game cities that can be used in winter). With the recommendation to establish tourist cities on the edges of the marshes, including the same components and infrastructure for water travel within the marshes, including (boats of different sizes + tour guides + maps of the owner's water within the marshes.
11- Recommending attention to the issue of periodic cleaning of the marshes and making the most of every water point in it and not wasting it and all other resources, by relying on modern technologies and working with specialized and efficient international companies.

Reference

[1] Al-Zahery, N., Pala, M., Battaglia, V., Grugni, V., Hamod, M. A., Kashani, B. H., ... & Semino, O. (2011). In search of the genetic footprints of Sumerians: a survey of Y-chromosome and mtDNA variation in the Marsh Arabs of Iraq. BMC evolutionary biology, 11(1), 1-16.
[2] Abboud.H. Najah, (1991). The Marshes of Iraq, an Environmental Study, Marine Science Publications, p.27
[3] Abboud.H.N, (1994), The Marshes of Iraq, an Environmental Study, Marine Science Center Publications, Basra, p34
[4] Abdul-Amir al-Hamdani and tagged , The Cultural Heritage of the Mesopotamian Marshes from the Sumerian to the Abbasid Era.
[5] Abdullah, S. Ibrahim, (2012). Earth Systems in Dhi Qar Governorate, Geomor Physiological Study, Master's Thesis (unpublished), College of Education, Al-Mustansiriya University .
Sources List:
[6] Abu Jarry, I. A. Hussein. (2007). Environmental Effects of Drying Marshes in Southern Iraq, PhD. Thesis, University of Baghdad, College of Education (Ibn Rushd), University of Baghdad, p.19
[7] Abdul-Razzaq.M, Thamer Salem Ali, (1994). The Importance of the Marshes in the Life of Some Species of Fish, The Marshes of Iraq, an Environmental Study, p 205-215.
[8] Abu Jarry, I. A. Hussein, (2007). Environmental Effects of Drying Marshes in Southern Iraq, PhD. Thesis, University of Baghdad, College of Education (Ibn Rushd), University of Baghdad, p.21
[9] Al-Aesawi, Q., Al-Nasrawi, A. K., Jones, B. G., & Yang, S. Q, (2021). Geomatic freshwater discharge estimations and their effect on saltwater intrusion in alluvial systems: a case study in Shatt Al-Arab estuary. Environmental Earth Sciences, 80(18), 1-15.
[10] Al-Nasrawi, A. K., Fuentes, I., & Al-Shammari, D, (2021). Changes in Mesopotamian Wetlands: Investigations Using Diverse Remote Sensing Datasets. Wetlands, 41(7), 1-17.

[11] Ali, M. H., & Al-Maliky, T. H. (2017). Fecundity of the crab, Potamon mesopotamicum Brandis, Storch & Turkay, 1998 from the Mesopotamian Marshlands, Iraq. Journal of Fisheries and Environment, 41(3), 6-11.
[12] Al-Khafaf, A.Ali, Al-Ziyadi, H. Alawi, Al-Fartousi, Kh. K., (2019). Al-Rafidain Center for Dialogue, Beirut Edition, Lebanon, Edition 1, Beirut, Lebanon, , p. 13.
[13] Al-Khafaf, A.Ali, Al-Ziyadi, H. Alawi, Al-Fartousi, Kh. K., Al-Fartousi, The Marshes of Iraq Three Studies in the Environment, Animals and Tourism, Al-Rafidain Resource Center, Edition 1, p.33
[14] Al-Khalili.M.J, tagged, (2012). developing tourism in the marshes and its impact on economic and social life).
[15] Al-Lata'i, Al-Qasab & Salim, Al-Jabayish Al-Ustad Magazine, Volume Ten Province in Southern Iraq, p. 256-255, , Part: 1, pp. 19-28.
[16] Al-Rubaie, A. A. Latif Ku, (2015). An Environmental and Morphological Study of the Marshes of Southern Iraq, published research, No. 23, 446-2008.
[17] Al-Tameemi, R,(2019). Societies Woven in Reeds: Reconstructing the Cultural Landscape of Nippur and the Iraqi Marshlands Through the Lens of John H. Haynes's Photographic Catalog (Doctoral dissertation, University of Cincinnati).
[18] Al-Tameemi, R, (2019). Societies Woven in Reeds: Reconstructing the Cultural Landscape of Nippur and the Iraqi Marshlands Through the Lens of John H. Haynes's Photographic Catalog (Doctoral dissertation, University of Cincinnati).
[19] Al-Zahery, N., Pala, M., Battaglia, V., Grugni, V., Hamod, M. A., Kashani, B. H., ... & Semino, O, (2011). In search of the genetic footprints of Sumerians: a survey of Y-chromosome and mtDNA variation in the Marsh Arabs of Iraq. BMC evolutionary biology, 11(1), 1-16.
[20] Al-Zaidi . K. Abdul-Amir M. (The Marshes of Southern Iraq between Sustainability and Challenges.
[21] Bushra Ramadan Yassin, Hussein Juban Oreib, (1999). Evaluation of Some Characteristics of Reclaimed Marshes Soils in Basra Governorate, published research, Iraqi Society Journal, Issue 39, Baghdad, p.32
[22] C, Mustafa Kazem BP, (2017). "Iraq's Ahwar: What does its inclusion in the World Heritage List mean". BBC Arabic. Archived from the original on February 02, 2019. Retrieved May 04.
[23] Food and Agriculture Organization of the United Nations (FAO),(2017). Iraq Country Profile. Available at:www.fao.org/ag/agp/AGPC/doc/Counprof/Iraq/Iraq.html, accessed 6 February .
[24] http://www.beatona.net/ar/knowledge-hub/article/content-40280
[25] Ihsan Mohsen Attia Al-Hafiri, Environmental pollution in the marshes of southern Iraq,
[26] Iqbal Abdul-Hussein Abu Jerry, (2007). Environmental Effects of Realizing the Marshes in Southern Iraq, Master's Thesis, College of Education (Ibn Rushd), University of Baghdad, p17.
[27] Khidher, S. A, (2019). Geographical Changes in the Marshes of Iraq. Journal of the College of Education for Women, 30 .
[28] Maatouq, S. Shaker, (2015). The hydrochemical changes of the southern Iraqi marshes and their environmental impacts, published research, Basra Studies Journal, eighth year, No. 15,P 238.
[29] Muhamad Ramdan, I., Candra, K. P., Lusiana, D., & Duma, K,. (2020). Redesign of the Traditional Handloom for Sarong Female-Weavers Based on Anthropometric Data.

Effluent Quality Assessment Monitoring of Al-Rustamiya Sewage Treatment Plant Using Geographical Information System (GIS)

Rusol Mohammed, Basim Hussein, and Athraa Hashim
MSc student University of Baghdad/Engineering college Baghdad and Iraq
Prof. Dr. University of Baghdad/Engineering college Baghdad and Iraq
Assist. prof University of Baghdad/Engineering college Baghdad and Iraq.

r.mohsin1901m@coeng.uobaghdad.edu.iq

Dr.basimal-obaidy@coeng.uobaghdad.edu.iq

Dr.athraa.h@coeng.uobaghdad.edu.iq

Abstract

Water quality assessment offers a scientific basis for water resource development and management. This research aims to assessment of Al-Rustamiya sewage treatment plant depending on annually changes and produces maps that declare changes on parameter during a period (2015-2018). Based on prior Government Department Baghdad Environment data which annually feature changes for samples from Northern Rustamiya have been estimated as a working model. Drawn a map of the Diyala River shows annual changes in the characteristics of the Diyala River, based on northern and southern Rustamiya effluent samples, and Diyala River samples. The characteristics that research focused on were biochemical oxygen demand (BOD), total suspension solids (TSS), chloride, cl, hydrogen ion concentration, pH, sulfate, SO_4^{-2}, nitrate, NO_3^- and phosphate, PO_4^{-3} during the period of (2015-2018). The results demonstrate that yearly variations in wastewater treatment characteristics occasionally violate Iraq's regulatory requirements. It was identified that the NO_3^- inadequate Iraqi standard for four years of study and SO_4^{-2} were inadequate for the first three years of the study period, TSS for two years outside the specification and cl and pH were compatible with period study. The river maps were separated into three groups where NO_3^- pollution from Northern Rustamiya with value 11.99 mg/L and TSS, Cl, SO_4^{-2} and PO_4^{-3} concentrated in centre of Northern and southern Rustamiya with values 96.99, 305,869.99, and 3.099 mg/L. The southern Northern Rustamiya area concentrates of BOD and pH with values 73.2 mg/L and 7.79. Effluents assessment of statically analysis and GIS for two Rustamiya stations provides a clear and integrated view of the output and impact on the Diyala river for Northern and Southern Rustamiya stations.

Keywords: Effluents Quality, Geographical Information System, Al-Rustamiya, and inverse weighted distance (IWD)

1. Introduction

As environmental concerns have grown, there has been greater attention paid to the appropriate functioning and supervision of wastewater treatment plants (WWTPs). The WWTP can generate major environmental and public health problems since the discharge of contaminated effluent to a receiving body of water might lead to various diseases and transmit them to humans and marine life. Environmental rules impose restrictions on effluent that each WWTP must adhere to[1]. Because there is difference of the pollutants concentration, the waste released into the river, it was found that these pollutants present in the river reduce the effects of solar energy absorption, resulting in a lower rate of photosynthesis and slows down natural water purification processes and the long effect of this is environmental degradation [2]. Sewage is any water that has been impacted by anthropogenic influences, as well as that disposed by domestic, industrial, commercial, and agricultural activities. It may also be contaminated by a wide range of pollutants and must be treated physically and chemically before being released into the water supply system or disposed of [3].

In other hand, much of the research focused on used modern technology in management water resources, one of this technology is GIS; is a very useful tool to develop solutions to water resource issues in order to assess water quality, determine water availability and understand the natural environment locally and/or regionally. The GIS makes it possible to map the spatial distribution of different pollutants. The resulting data is real useful for decision-makers to consider remedial steps [3]. One of the most helpful instruments for the planning, management and protection of the wastewater environment is updated spatially-oriented data system Consequently, a system providing geospatial connections with facilities needs to be provided. One of the key components is to find wastewater treatment systems and to collect, analyze and visualize geographically oriented data. Furthermore, GIS-BASED decision support systems for the treatment plants available are well-suited to inform better analyzes and understanding of the problems of the available processing facility to improve the analysis and understanding of the current situation in choosing the appropriate treatment location using ArcGIS. [4]

There has also been a study paper investigating the application of remote sensing and GIS in two major locations (Al Ballanah and Al Allaqi) in Aswan, Egypt for evaluating the hydrological difficulties of wastewater. In general, wastewater effluent is supplied through pressure lines to secondary oxidation ponds, to monitor surplus wastewater accumulation, Hydrologically develop a water balance model to be used to manage such regions and to designate in the vicinity alternative locations appropriate for water storage and agriculture, therefore reducing adverse environmental effects of wastewater [5].Assess the risk assessment of Olive Mill Waste Water Greece has been studying spatially and temporally on the basis of GIS technology in Crete in terms of the control elements linked with it. These control parameters are vital to the modeling and simulation of the process of risk generation. In addition, a complete overview was taken of the probable sources of pollution, predicted receptors, as well as the paths that connect them. In every sub catch in the analyzed watercourse, a computation model was developed to objectively evaluate the danger. In order to compute the global possible danger a catch may suffer due to the contribution of all pollution sources within the watershed. It has been built for this purpose [6].

The objective of this research is to assessment monitoring of Al-Rustamiya sewage treatment plant depending on yearly changing and produce maps that declare changing on parameter during (2015-2018).

2.Methodology of work

The annually changes in characteristics have been computed for samples supplied by the North Rustamiya as working model, based on the prior data received from the test laboratory from the Baghdad environment, which can be found later in the results section. As for the characteristics that the research focused on were biochemical oxygen demand (BOD), total suspension solids (TSS), chloride, cl, hydrogen ion concentration, pH, sulfate, SO4-2, nitrate, NO3- and phosphate, PO4-3.

Satellite image that obtained from Amant Baghdad for case study area used as a base to produce maps. Geographical Information System(GIS) program (ArcMap 10.7) used and based on the data of effluent samples presented from the northern and southern Rustamiya and the samples examined from the Diyala River relying on the analyser spatial interpolation tool and inverse weighted distance (IDW) method was adopted to produced maps that presented variation of parameters.*2.2. Fied operationa*
The experiment included three blocks by reality 9 experimental units for each block and dimensions (2 × 2) m, the experimental units contained ten agricultural lines and a distance 20 cm between the lines. variety Bohouth 22 seeds were sown 15 / 11 / 2019 at a rate 120 Kg. ha^{-1} Mineral fertilizer add for (N, P) by 120 Kg N .ha^{-1} , 100 Kg P ha^{-1} [9]. Mineral and nano fertilizer section for three growth stages (ZGs21 Tillering, ZGs32 Stem Elongation, ZGs49 Boot). Harvest the crop on a date 20/4/2020.

2.1.Study case description

The city of Bagdad has around 900 square kilometers and approximately 7.6 million people for the year 2010. The city is quite vast and virtually flat split into two main portions by the Tigris River: the east (Rusafa) and the west (Karkh). The city consists of 457 sectors with sewage systems serving roughly 82 percent of the sectors. Baghdad city has three projects for wastewater treatment. These projects are Al-Rustamiya South Station, AL-Rustamiya North Station and Al-Karkh [7]. The Al-Rustamiya STP is Iraq's biggest sewage treatment plant, located on the southern side of al-Rusafa in Bagdad, about 500 meters from the Diyala River, and has a 400 000 m^2 area, as indicated in Figure 1. [8]. The Zeplin line is the raw sewage input line to the plant. That collects Rusafa sewage at design capacity and actual flow rates of 300,000 and 450,000 m^3/day, respectively. It is split into two independent outputs (F1 and F2). The sewage is processed at different stages inside the plant before being discharged into the

IOP Conf. Series: Earth and Environmental Science 1002 (2022) 012011 doi:10.1088/1755-1315/1002/1/012011

Diyala River. The Diyala River is one of Iraq's most significant rivers, with a flow rate ranging from 25 to 650 m^3/s [9]. This plant's treatment stages include screens, grit chambers, aeration tanks, primary sedimentation tanks, secondary sedimentation tanks, chlorination, and sludge treatment, which process the raw sewage physically, chemically, and biologically to meet Iraqi regulations to discharge effluents of STP to the river [10].

Rustumiyia WWTPs are oldest wastewater projects in Iraq, providing public health promotion services to about one third of the population of Baghdad to drain treated sewage into the Diyala River. The project is located at the side of Al-Rusafa in the south-eastern part of Sakina and Kiryat provinces near the New Diyala Bridge, and is about 500 m from the Diyala River and the nearest residential community about 2 km As for the nature of the land built on it, it is agricultural and residential and the prevailing wind trend is with the direction of the housing units and the prevailing wind trend is with the direction of the housing units, which made it not identical to the environmental determinants of the establishment of such activity [11].

Figure 1. Rustamiya WWTP location in Baghdad (Researcher's work).

Figure 2. Rustamiya WWTP location in Baghdad (Researcher's work).

TABLE 1. *The stages of treatment, the year of construction and the design capacity of the Rustamiya plant [11].*

Name of stages	*Year of construction*	*Design capacity (m3/d)*	*Serviced population (capita)*
Zero stage (F0)	*1963*	*40,000*	*350,000*
First Expansion(F1)	*1975*	*45,000*	*400,000*
Second Expansion (F2)	*1985*	*90,000*	*750,000*
Third Expansion (F3)	*1988*	*300,000*	*1,500,000*

TABLE 2. *capacity of the Rustamiya plant [11].*

Project	***design (m3/d)***	***Received (m3/d))***	***Treated (m3/d)***
South Rustumiyia	*250,000 (75,000+175000)*	*350,000*	*250,000*
North Rustumiyia	*300,000*	*450,000*	*300,000*
Discharge of waste water received from Rusafa (m3/d)	*(South Rustumiyia+ five compact+ North Rustumiyia)=800 m^3/d*		

2.1.1. Northern Rustamiya

The area of project is about (400,000) m^2 .it designed from two Parallel and equal treatment lines [11].

Figure 3. Locations of North Rustumiyia WWTP in Baghdad (Researcher's work).

2.1.2. Southern Rustamiya

South Rustumiyia Project established in 1963, is the oldest A project in Iraq where it was implemented by the Company (Azmal) Indian company with the supervision and design of the entity Contracted English. The project extends over an area of 700,000 m^2 equivalent to about 280 dunums [12].

Figure 4. Locations of South Rustumiyia WWTP in Baghdad (Researcher's work) .

2.1.3 Diyala River

Diyala River originates from the highlands of Iran and Iraq, with an altitude of more than 2,000 meters. The overall distance from the Diyala River is 386 km, around 300 km of which are inside the boundaries of the Province of Diyala and flow into the Tigris River, approximately 32 km south of Baghdad. The Diyala Basin is about 31,896 km^2 with an area of 24,072 km^2 within Iraqi borders and an average annual water flow of approx. 5,8 billion cubic meters that before project on this River was completed by Iran. It is worth noting that Iran is regarded as an upstream nation with regard to Diyala River's geographical position. Iraq is considered a downstream country, on the other hand. [13].

Diyala River, which rises in Zagros Mountains in Western Iran, is the most significant eastern affluent of Tigris River. It covers distance of 445 km of which 386 km in Iraq (Fig 5). The catchment can be split into 3 parts. The upper Diyala stretches from its sources into the Dam of Derbendekhan, the middle Diyala extends from the Derbendechhan downstream, the Himreen dam area and the lower Diyala area from the submerged Dam of Diyala downstream to the 10 km south of Baghdad, with a mean steepness of about 1-85 m/km Before reaching the Tigris River, there are two wastewater treatment plants in the Diyala River. Tangro, Servin and Wand, with a discharge of 181 x 103 / day, are the most major branches of the river. Its area about 32 x 103 km^2 [14].

Figure 5. Location of Diyala River (Researcher's work).

2.2 *Geographic information system (GIS)*

Water management has become the focus of the global study on increasing awareness and concern. The establishment of surveillance networks in Maine has led to a vast database of time and space changes in pollutant pollution. In contrast, most of the Mediterranean nations (including Greece) have no regular quality monitoring, although many pollution concerns have been identified. A complete assessment of environmental issues on the riverside scale has to be made in order to provide a more suitable scientific foundation for controlling the consequences of human activity in sustainable management of river systems. [15].

Arc GIS Geo-processing is a key component of ArcGIS and comprises various procedures which may be used to extract data. The study highlighted the relevance of contemporary technology to address wastewater treatment plant environmental concerns (WWTP) in advancing environmental protection. The research approach used remote sensing and GIS tools to make the study easier and offer relevant information about the field. It also gives the standard ways with additionally proper outcomes. GIS has therefore shown to be a strong instrument in the field of relevant analysis for the particular non-space information management [16].

The inverse weighting or inverse weighted distance technique estimates at unprocessed places an attribute value, using a linear combination of points at the sampling points weighted with an inverse function of the distance from the place of interest to the sampled locations. The presumption is that the sampled items nearer to the location that was not sampled are more comparable than those farther away. Weights can be shown as:

$$\lambda_i = \frac{1/d_i^p}{\sum_{i=1}^{n} 1/d_i^p} \qquad 1)$$

Where di is the distance from x0 to xi, p is the power parameter, while n is the number of points to be sampled for this estimate. The key component that affects IDW's accuracy is the power parameter setting. Weights decrease as the distance grows, especially as the power parameter value increases, so that close samples have a higher weight and impact the estimate more and spatial interpolations are local. The power and quarter size decision is random.

P 2 is the most prevalent alternative, and this approach is commonly referred to as reverse square or reverse square distance (IDS). On the basis of error measurement, too, the power parameter may be set for the smallest average absolute error, resulting in optimal IDW. As the power parameter grows, the smoothness of the estimated surface rises, and it was discovered that the estimated results are less satisfying when p is 1and 2 compared to p is 4. IWD shall be referred to as "moving average" if p is null, when p is 1, and when p is not equal to 1. IWD shall be called "moving average" [17]

When a group of points is dense enough to record the degree of local surface variation required for analysis, the IWD feature should be employed. IWD uses a liner weighted collection of sample points as a means of determining the cell value. The allotted weight depends on the distance from the output cell position of a point to be entered. The bigger the distance, the less cell impact on the output value [18].

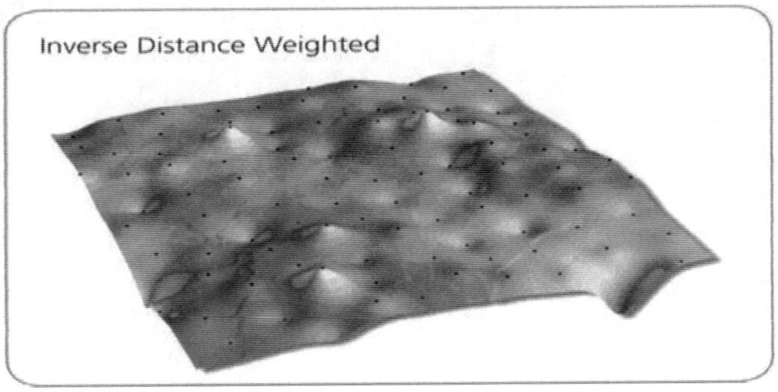

Figure 6. Model of IDW feature surface [14].

3. Results and discussions

3.1.Effluent quality using assessment

By statistically evaluating data from laboratory testing of treated sewage that discharged into the Diyala River during 2015–2018, it was observed that the maximum BOD and TSS were recorded in 2016, which did not indicate Iraqi requirements (IOMOH, 2012) as shown in Fig. 7 and Fig. 8. The pH parameter denotes the acidity or alkalinity of water. In 2017, the greatest values of development of PH were reported in accordance with Iraqi requirements (IOMOH, 2012), as well as for other years in the period (2015–2018), which are within the limitations of Iraqi requirements as shown in Fig.9.

When compared to Iraqi regulations, the amount of chloride in the effluent in disposals was appropriate. The year with the highest value was 2015, while the year with the lowest value was 2016. (Fig.10). The change in nitrate (NO_3) during four years (2015–2018) was extremely excellent when compared to the limitations of Iraqi regulations for treated wastewater disposal in rivers. However, phosphate was outside of Iraqi standards in 2015 and 2018. (Fig. 11 and Fig. 12). The Sulfate (SO_4) variation throughout four years (2015–2018) exceeded Iraqi standards for three years, with the exception of 2018. Additionally, SO_4 levels greater than 200 mg/L hampered sludge digesters (Fig. 13).

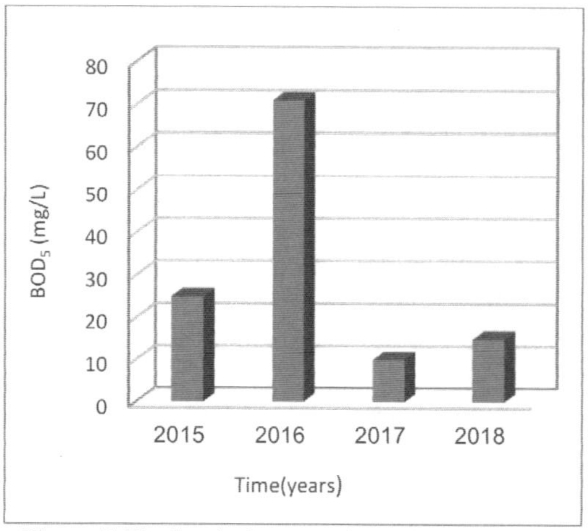

Figure 7. Biochemical oxygen demand (BOD) annually variation during (2015-2018).

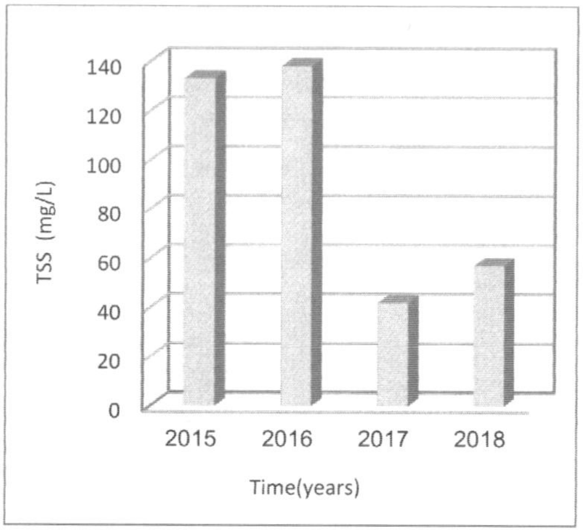

Figure 8. Total suspension solids (TSS) annually variation during (2015-2018).

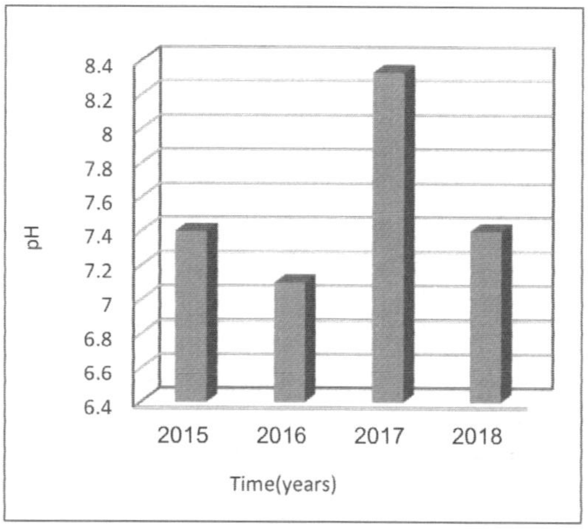

Figure 9. pH annually variation during (2015-2018).

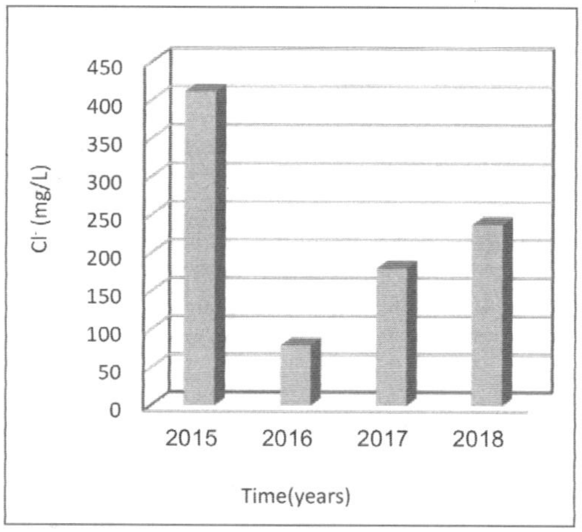

Figure 10. Chloride (Cl^-) annually variation during (2015-2018).

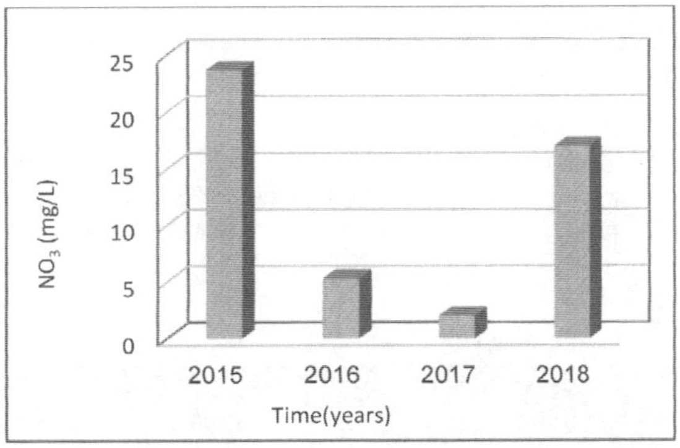

Figure 11. Nitrates (NO_3^-) annually variation during (2015-2018).

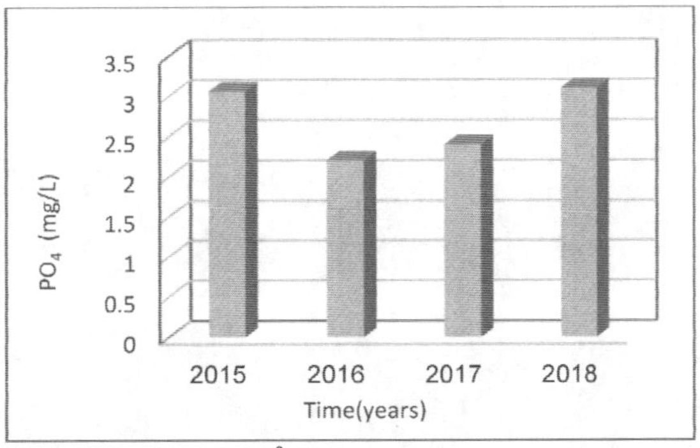

Figure 12. Phosphate (PO_4^{-3}) annually variation during (2015-2018).

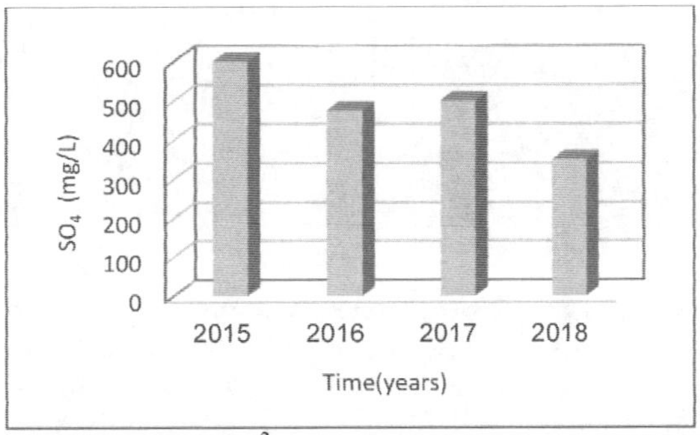

Figure 13. Sulphate(SO_4^{-2}) annually variation during (2015-2018).

3.2. Effluent quality assessment using GIS

Based on laboratory test results for the outputs from both Rustamiya stations and from the Diyala River samples, the program GIS using an IWD method used to obtain maps showing the river division into areas where the maximum pollution area in southern Rustamiya for BOD and pH (Fig.14 and Fig.15) with values 73.2 mg/L and 7.79, while the pollution area was concentrated in the area between the northern and southern Rustamiya for TSS, cl, SO_4^{-2} and PO_4^{-3} (Fig.16, Fig.17, Fig.19 and Fig.20) with values 96.99, 305,869.99,and 3.099 mg/L, As for the area near the northern Rustamiya, NO_3^- concentrated in it (Fig.18) with value 11.99 mg/L.

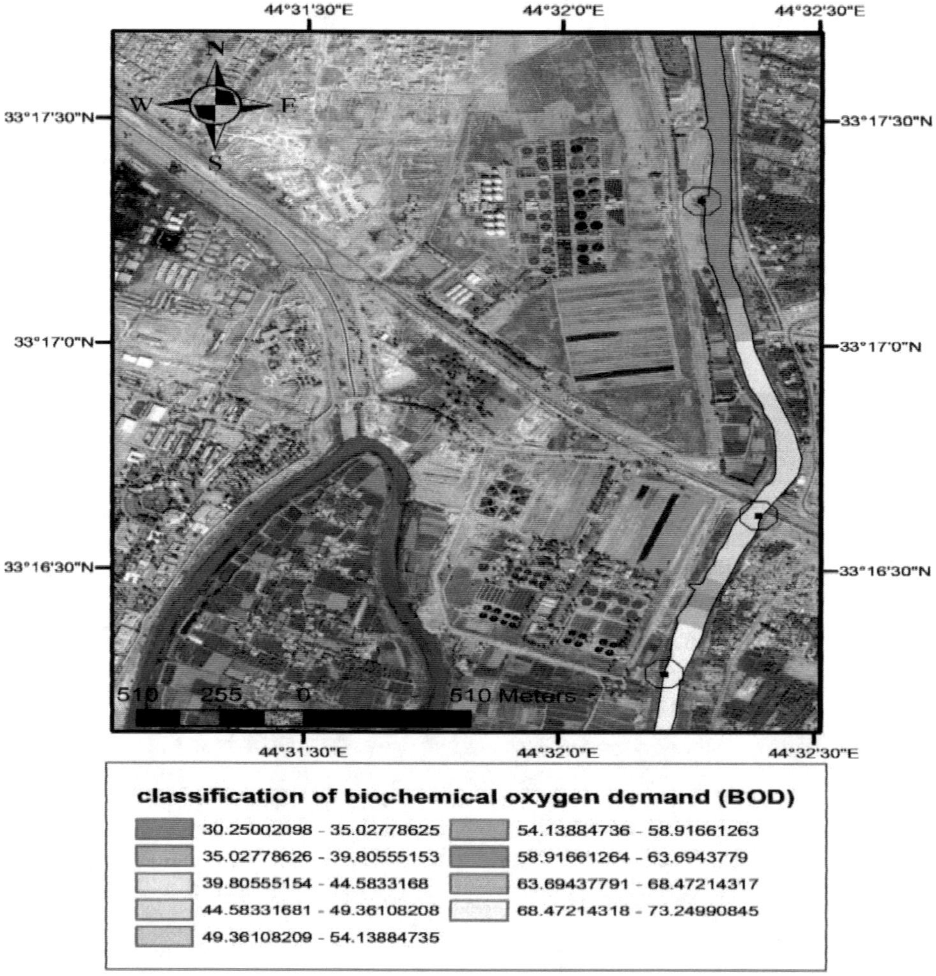

Figure 14. Biochemical oxygen demand (BOD) annuall variation y during (2015-2018).

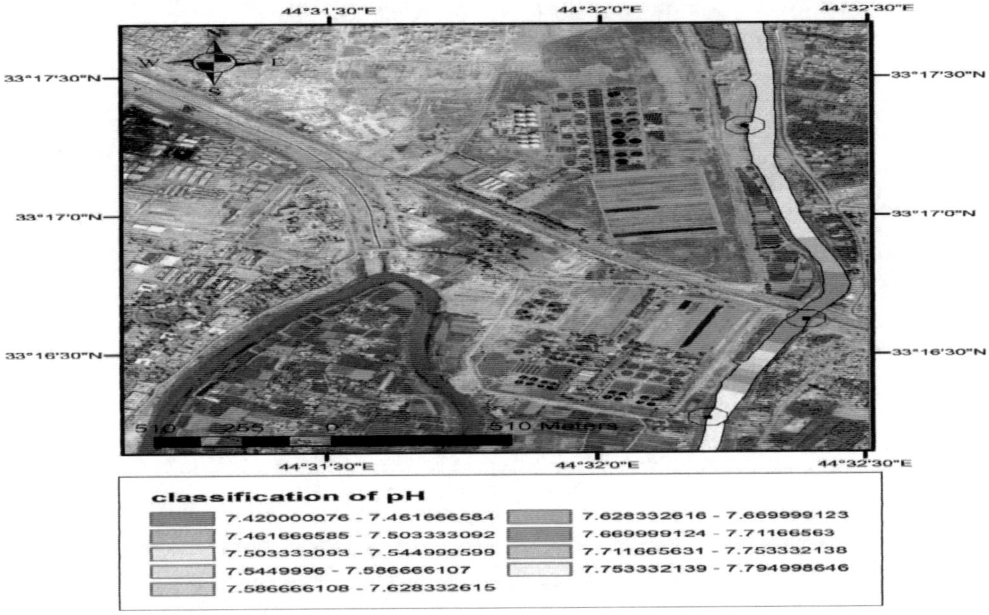

Figure 15. pH annually variation during (2015-2018).

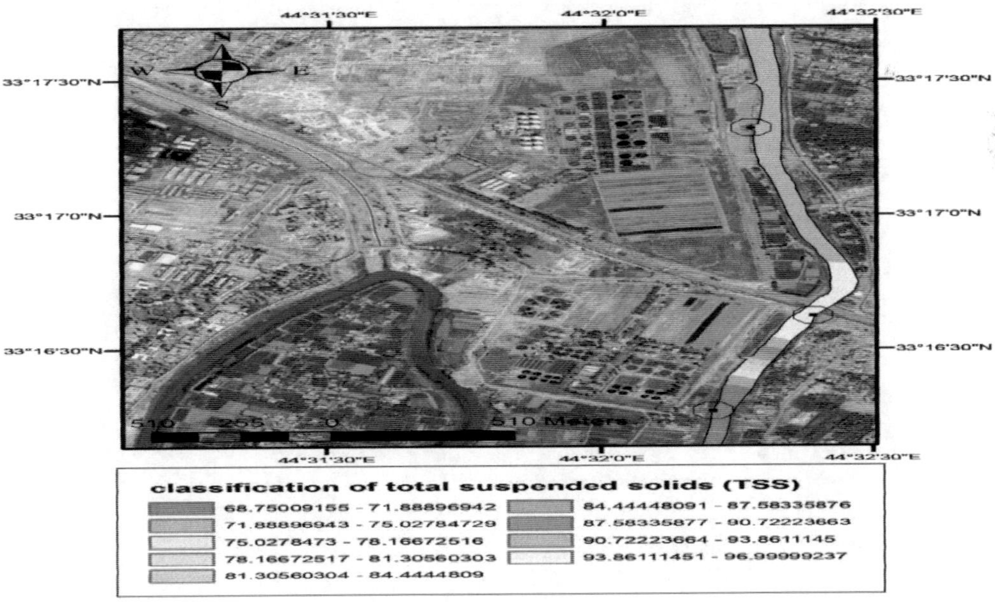

Figure 16. Total suspension solids (TSS) annually variation during (2015-2018).

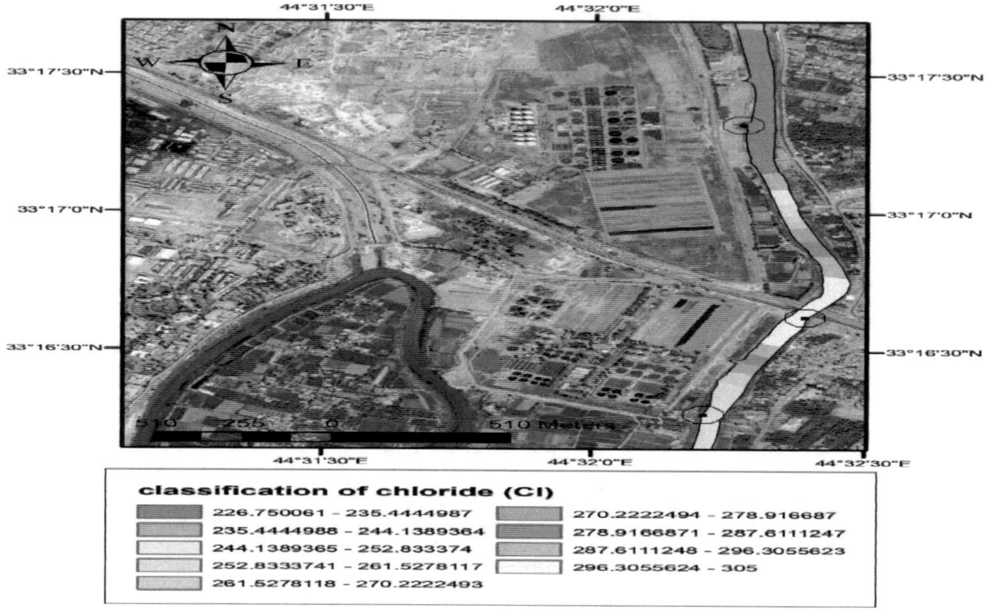

Figure 17. Chloride (Cl^-) annually variation during (2015-2018).

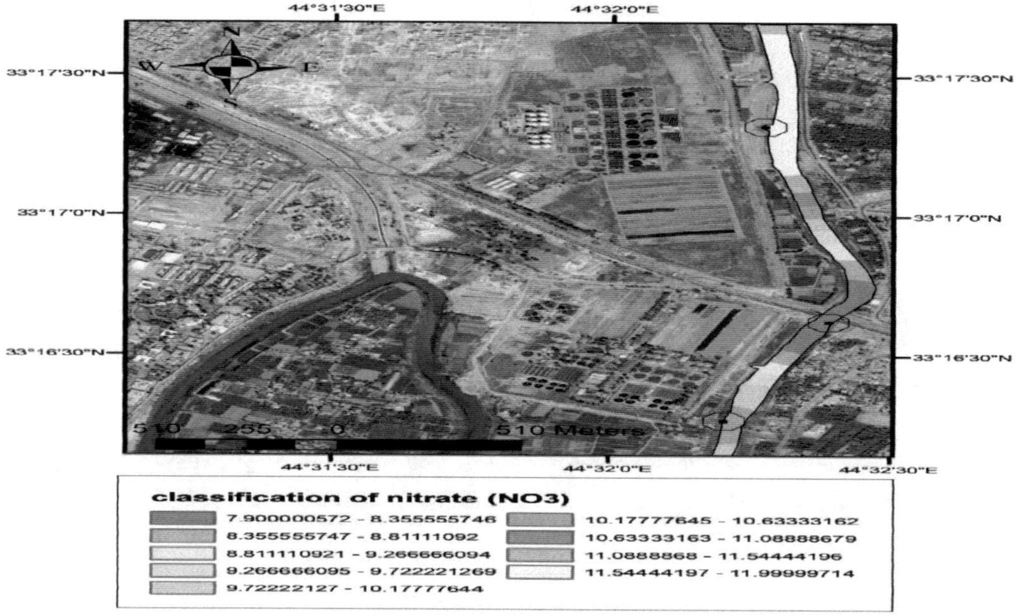

Figure 18. Nitrates (NO_3^-) annually variation during (2015-2018).

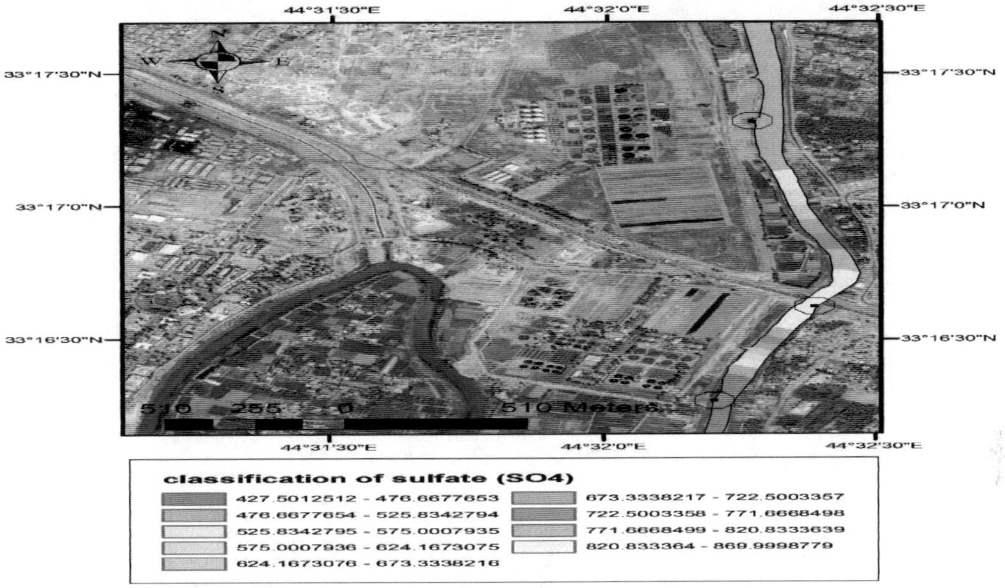

Figure 19. Sulphate (SO_4^{-2}) annually variation during (2015-2018).

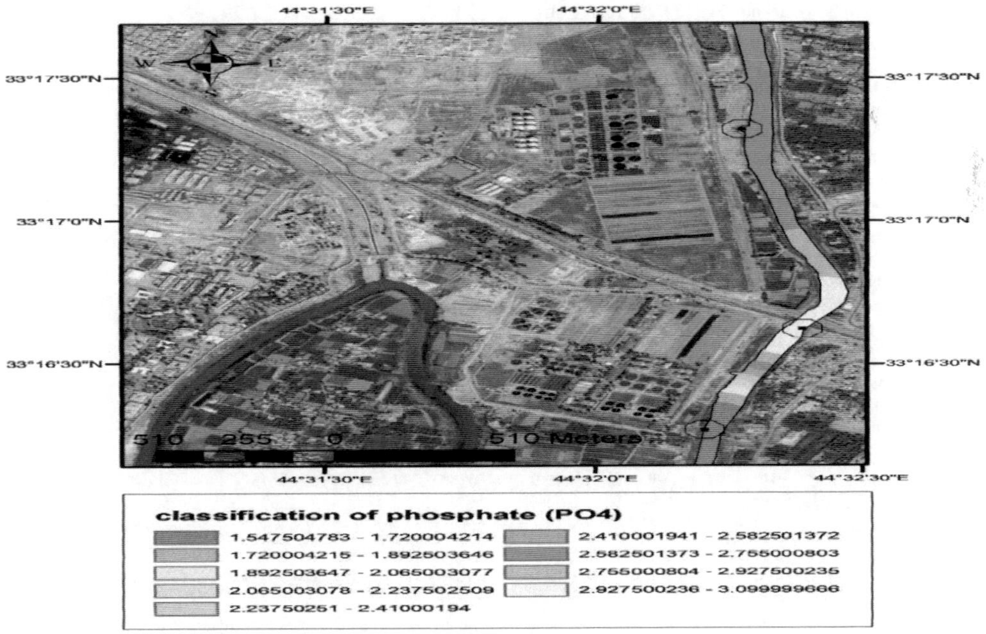

Figure 20. Phosphate (PO_4^{-3}) annually variation during (2015-2018).

4. Conclusions

The following are the major results that has been achieved:

1- The annually changes of features and respect of Iraq's standards for wastewater treated effluent may be understood by laboratory testing.

2- SO_4^{-2} rates were not identical to the Iraqi specifications in the first three years of the study. Two-year TSS was out of specification, but Cl and pH were compliant through the years of that study.

3- Maps split into three regions in which Northern Rustamiya concentrated the greatest NO_3^- pollution levels, and between northern and southern Rustamiya the TSS, cl, SO_4^{-2} and PO_4^{-3} concentrated. BOD and pH concentration in Rustamiya South.

4- The GIS program was used to build data bases that created maps showing the pollution and division of the river into several changing zones.

5- Assessment effluents for two Rustamiya stations statically and using GIS gives a clear and integrated picture of the outputs of the submissions of the two Rustamiya stations and their impact on the Diyala River, this paves the path for future research to be more successful and sophisticated.

References

[1] **Mohammed, R., and Al-Obaidi, B, 2021**, Treatability influence of municipal sewage effluent on surface water quality assessment based on Nemerow pollution index using an artificial neural network. In IOP Conference Series: Earth and Environmental Science, Vol. 877, No. 1, p. 012008. IOP Publishing.
[2] **Lekwot V E; Caleb A I and Ndahiayiem A K, 2012**.Effects of effluent discharge of Kaduna refinery on the water quality of river Romi. Journal of Research in Environmental Science and Toxicology.
[3] **Erabee, I. K., & Ethaib, S. J. I. J. ,2018.** Performane of Activated Carbon Adsorption in Removing of Organic Pollutants from River Water. International Journal of Engineering & Technology, 7(4.20), 356-358.
[4] **Alanbari M ; Alquzweeni S S and Aldaher R R,2015**. Spatial distribution mapping for various pollutants of al-kufa river using geographical information system (GIS). International Journal of Civil Engineering and Technology (IJCIET).
[5] **M.A G and Ebraheem J , 2012.** Location Optimization of Wastewater Treatment Plants Using GIS: A Case Study in Umm Durman/Karary. Annual Conference of Postgraduate Studies and Scientific Research (Basic and Engineering Studies Board).
[6] **El Bastawesy M; Shalaby A and mohamed I. 2017**. Assessment of the hydrological and environmental problems of waste water disposal sites in aswan of egypt using remote sensing and GIS. International Journal of Hydrology.
[7] **Elhag M; Al-Ghamdi A.A.M; Bahrawi J.A and Hegazy A.K. 2017**. Olive Mill Waste Water Risk Assessment Based on GIS Techniques in Crete, Greece. Global NEST Journal.
[8] **AbdulRazzak A. M.,2013,** Performance Evaluation of Al-Rustamiya Wastewater Treatment Plant. Journal of Engineering, 19, 429-438.
[9] **Al-sakini ,2013.** Journal of KerbalaUniversity 11 195-203
[10] **Al-obaidi B H K, 2020** Journal of Engineering Science and Technology 15 3571 – 3587
[11] **Ismail H A ,2016** Journal of Basra 76 243-267.
[12] **Al-sakini, A.,2016**, Rustumiyia sanitation project and its effects on polluting Diyala river. Basra Literature Magazine,76, 243-267.
[13] **Al-Dulaimi, Muhammad Abdul-Razzaq Mahdi** "Physical and chemical properties of project water Rustumiya sewage and treatment efficiency, Master's thesis, University of Baghdad, College of Education for Pure Sciences, Ibn Al-Haytham, Department of Life Sciences, 2015.
[14] **Hussein, O., Masum,A. and Mohamed K., 2016**, The Use of the Diyala River Between Iran and Iraqi: An International Law Approach, Journal of Legal and Political Sciences,5(1),285-305.
[15] **Hadi, R. A, Ismail, A. M. and Talib, A. H.,2009**, Check List of the algae in Diyala River, Iraq, Um-Salama Science Journal, 6(2), 923-345.
[16] **Dassenakis M; Scoullos M; Foufa E; Krasakopoulou E; Pavlidou A and Kloukiniotou M,1998**. Effects of multiple source pollution on a small Mediterranean river. Pergamon.
[17] **Abbasl Z d and Jassim O, 2019**. Application of GIS and AHP Technologies to Support of Selecting a Suitable Site for Wastewater Sewage Plant in Al Kufa City . Al-Qadisiyah Journal for Engineering Sciences.

[18] **Jin Li and Andrew D. Heap,2008**, A Review of Spatial Interpolation Methods for Environmental Scientists, Geoscience Australia, GPO Box 378, Canberra, ACT 2601, Australia,
[19] **colin childs, ESRI education services,2004**, ARCUser, interpolating surfaces in ArcGIS spatial Analyst.Strite S and Morkoc H 1992 *J. Vac. Sci. Technol.* B 10 1237.

Acknowledgments
The authors are grateful to the environment Baghdad government institution for providing sufficient chance to collect data and communicate with the station manager. The authors will indeed appreciate the sanitary engineering laboratory team as well as the Civil Engineering/ University of Baghdad for their invaluable assistance in completing this study.

Use Remote Sensing Techniques to Study Epiphytic Algae on *Phragmites australis* in Um El-Naaj Lake, Mysan Province, Southern Iraq

AL NAQEEB, Neran A.[1&3*]; **AL HASSANY, Jinan S.**[2]; **MASHEE, Fouad K.**[3];

[1]Department of Biology, College of Science, University of Misan, Maysan, Iraq
[2]Biology Department, College of Science for Women, University of Baghdad, Baghdad, Iraq
[3]Remote Sensing Unit, College of Science, University of Baghdad, Baghdad, Iraq

* Correspondence author e-mail: neranecology@uomisan.edu.iq

Abstract

The aim of this study was to determine the quality and quantity composition of epiphytic algae of *Phragmites australis* in Um El-Naaj lake . Samples were collected from six Stations from November 2018 to October 2019.

During this study , a remote sensing techniques was used to monitor the Community of epiphytic algae through storage and management data the geographic information system (GIS) program. The satellite images of the Landsat 8 satellite of the OLI which used to create visual images that facilitate the interpretation of the study area to identification and presence of these species and control their spread.
The total number 306 species of epiphytic algae recorded during the study period were, belong to 61 genera. Seasonal variations in total number of species also noticed during the current study, Bacillariophyceae recorded 52.28%, with (160 species of 28 genera) , Classes of Chlorophyceae and Cyanophycea recorded 12.67% (39 species of 20 genera 11) and 11.76% (36 species of 12 genera) respectively , while Classes of Fragilarphyceae recorded 14.37% (44species of 7genera), Coscinodisphyeceae recorded 13.59% (11 species , 7 genera). Then class of Euglenophyceae with 3.59%. , Finally Chrysophyceae, Dinophyceae, and Rhodophyceae recorded 0.32% with 1 species of 1genera as they were present.

Keywords: *Phragmites australis,* Um El-Naaj marsh ,Remote Sensing Techniques, Epiphytic algae

1. Introduction

The large and basic Iraqi marshes are concentrated in southern Iraq, which are natural areas that were formed during the ancient sedimentary tectonic eras and plate fractures by moving processes inside the ground, which made the rivers and depressions of these areas take the natural form for thousands of years past[1], The nature of the marsh areas and the huge quantities of water as storage contribute to the natural balance of life on Earth and the preservation of land cover and life. These marshes feed into the main Iraqi rivers, the Tigris and Euphrates [2].
The rivers coming from neighboring countries such as Iran are considered natural reservoirs of escaping waters, floods and heavy rains, which helped these features to form a natural balance for the revival and hydrological cycle of the land cover for 5000 years ago [3].
The study area (Al-Hawizeh Marsh) is located between latitudes 31 and 31.75 north. Its waters are shared between Iraq and Iran and are not separated by borders, and the greater part of the water lies within the borders of Iraq and its total area is estimated (156.850986 km2). And a width of 30 km [4].

Epiphytic algae, including freshwater angiosperms and macroalgae, are a group of algae found attached to and living on submerged aquatic vegetation [5.] In the littoral region, these species are known to be the primary food source for small fish and other invertebrates.
Due to the repetitive process of separation of epiphytons from the host plant [6], these autotrophs are frequently ignored, Recent study, however, has shown its importance as primary producers that absorb essential nutrients and fix carbon from the water column, making these critical nutrients available to other species such as small aquatic habitats in the lakes [7]. Often, epiphytic algae in an aquatic ecosystem has serve as strong measures of water quality and environmental conditions. As they are sessile in nature, have short generation times and each species has its own collection of environmental tolerances and preferences, these organisms create an ideal bioindicator [8].
In addition, these organisms in the lotic water system are the dominant species and play an important role in the ecological equilibrium between several forms of macrophytes and their respective water bodies [9].

2. Materials and methods

Six sites were selected in Hor Al Huweiza, Um El-Naaj Lake as shown in(figure 1); identified by GPS as in(table 1).
The study was conducted from November 2018-October 2019. quantitave study includes after separate algae from plant putting the sample in cylinders put Lugol's solution for precipitation, Marked containers sample with the date and site collection. to identification epiphytic algae preparing permanent slide examination under the microscope. [10].

2.1. Study area

Um El-Naaj Lake, one of the Iraqi marshes, which is located in Kahala district, and has two wiled entryways, through AL Kahla district - in the direction of Bani Hashem – Um El-Naaj Lake, and another entrance is Kahla – Al mueel-Abo Khasaf village. Um El-Naaj Lake Up to25 km in width and 30km depth to the Iranian borders. He was part of the partially dried marshe, then returned to him naturally after the fall of Saddam Hussein's regime to flood a large area of it, but still there dried lands. Currently, this marsh is fed and revitalized with water through its outlets from inside Iraq like Alekhala River, AL Mshrah River and -outside of Iraqi like Aldwiridj and Karkheh, Nissan and Alkfagia [11].

Table 1. Points of coordinates of sites of monitoring stations and sample collection in the study area.

Sites	The coordinates	
	Longitude (eastwards)	Latitudes (northwards)
ST1	47°36'117"	31°37'022"
ST2	47°38'767"	37°38'409"
ST3	47°38'662"	31°38'028"
ST4	47°34'824"	31°37'143"
ST5	47°39'272"	31°37'370"
ST6	47°40'79"	31°36'887"

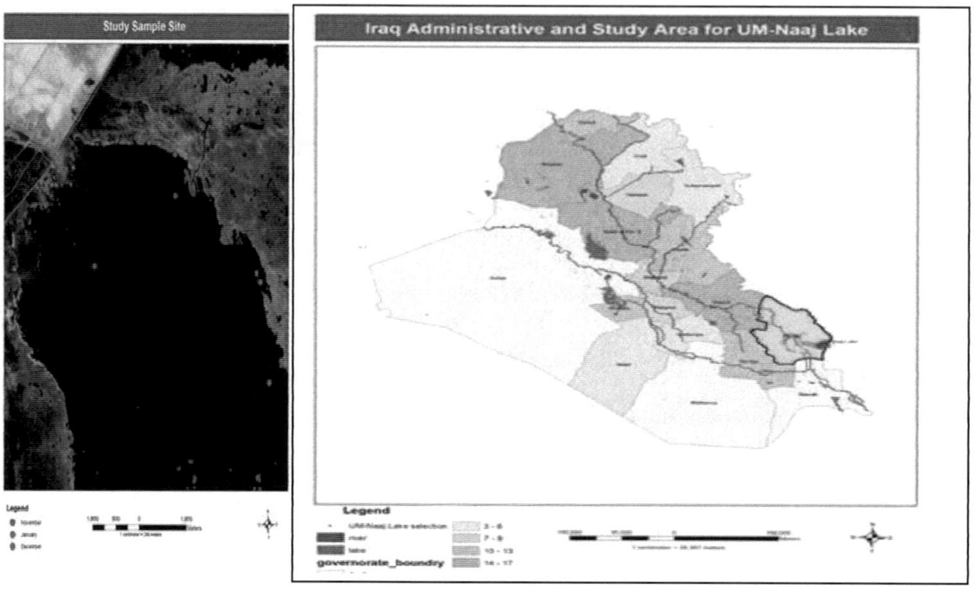

Figure 1. The study was conducted……., Study Area Site in Missan city, Observation station name in study area Al-Naaj lake include six stations .

2.2. Remote Sensing (RS) Processing.

In our study, we used four bands of terrestrial remote sensing data Landsat-8 OLI, for track: Column 167 and Row: 38, was produced in November 2018 through October 2019, which simultaneous with the timing of fieldwork data collection, images were downloaded from the WEB [12, 13].

The data of Bands 7, 4 and 3 were to compose in the process of finding the best clarity of vision for the study area, and to monitor the changes occurring from analyzing the parameters samples from the six stations spread over the study area [14]. The engineering correction was done to adjust these ground coordinates by the method of ground control points (GCPs) with the local basic map data for the study area.

These visual images allow us to provide a good interpretation of control, monitoring and decision-making in the course of research.

Using the features of converting Landsat satellite images first to higher atmospheric radiation (TOA) as radiation calibration parameters in the metadata file, this feature allows finding research features as a spectral comparison with laboratory results [15].

ArcGIS 10.5 is the digital image processing program used to process Landsat 8 images. Under which ground control points (GCPs) are determined from the images and corrected to UTM (Universal Transverse Mercator). Projection of the 1984 World Geodetic System (WGS 84) using coordinates obtained from the Global Positioning System (GPS) [16].

For the better interpretation of the analyzed results The concentrations of various parameters collected at different locations were integrated with GIS for a better understanding of the analyzed findings. The field of research is delineated from date Landsat Satellite 8 OLI sensor data . For the preparation of spatial distribution, the spatial database developed from the analysis of samples has been integrated with GIS [17].

3. Results and Discussion

The Seasonal Variations in the Total Number of Epiphytic Algae Species by interpolation of Inverse Distance Weighted.

The study covered two parts, part for the quantitative study of algae, and part two, the creation of contour and interpolation maps for the quantitative study of algae in the Geographic Information System, Um El-Naaj Lake in the Hawizah Marsh for a period of four seasons.

Bacillariophyceae recoded the larger number in autumn 125.3× 10^4 cell/gm site 2 these at the pink color which map while chlorophyceae recorded the highest value 23.9× 10^4 cell/gm site 5 represent the color pink in the map the lowest number 8.9× 10^4 cell/gm site4 green color
It is possible to determine the highest and lowest quantities, and the boundaries between them are inserted through colors or contour lines, through which we can sense the limits of this type in the future, and this method has become a predictive tool in our study.
The seasonal shift results obtained from evaluating reflectance values (value per sampling point) were then used as input data in ArcGIS 10.2. The sampling locations for the generation of spatial distribution maps are integrated with the water data. The current research uses the method of Inverse Distance Weighted (IDW) [18] for spatial interpolation of water reflectance in the marshes of AL-Hawizah 2018-2019, as shown (fig 2) [19].
This gave us real and expected values without reaching them or taking readings from them. This is an advantage and benefits of this technology and Each color has the same true value without reaching this area.
We note a decrease in the percentage of algae in the fall season because the beginning of the rainy season and the marsh was suffering from drought during the previous seasons [20]. And low nutrient concentration, increasing turbidity that decrease penetration light in the water column that cause incomplete photosynthesis of diatom and exposure diatoms to graze from fish and invertebrate cause reduction diatom numbers [21].

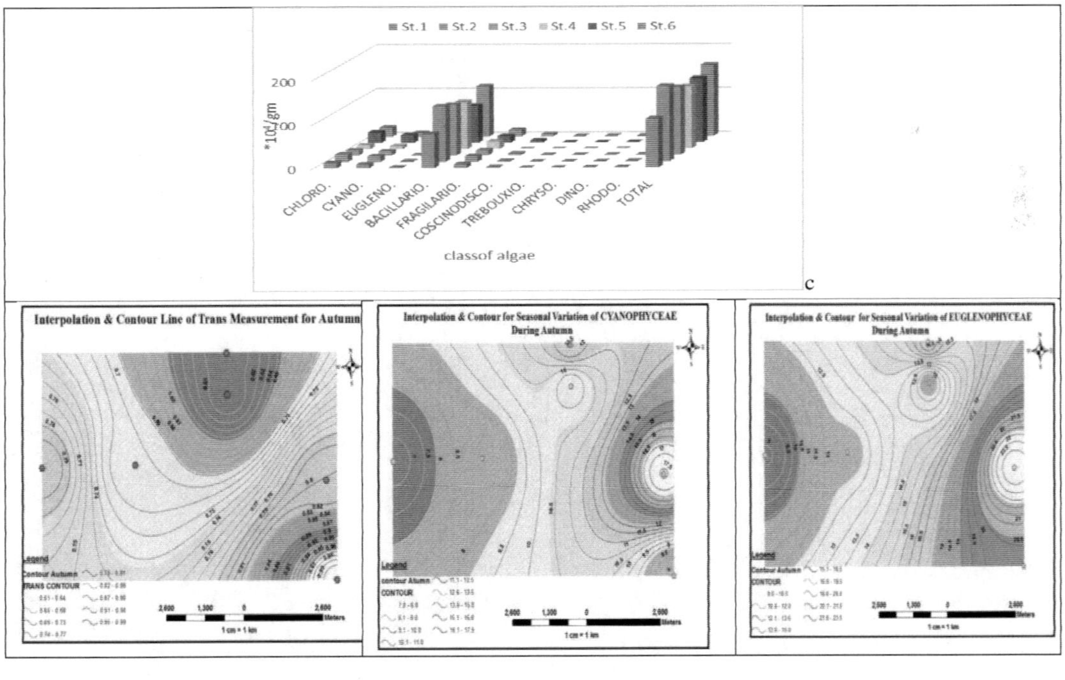

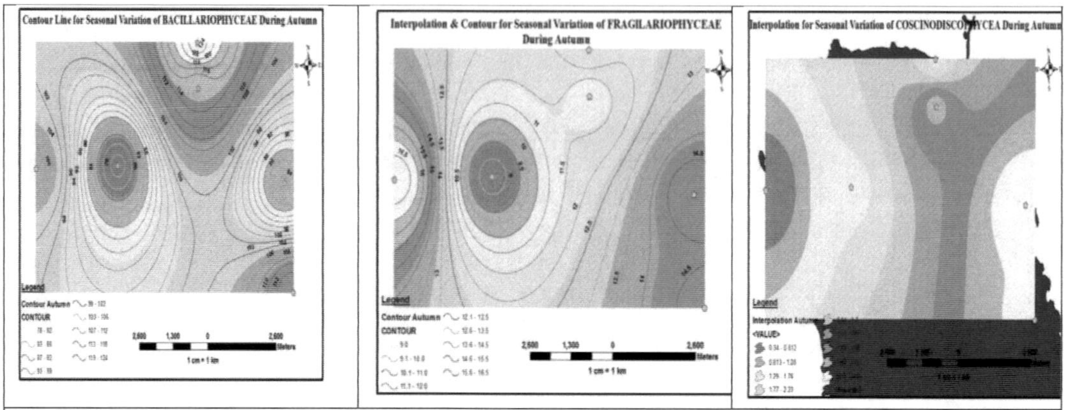

Fig 2. The interpolation of Inverse Distance Weighted (IDW) and Contour Lines techniques process determining the quantities.

Chlorophyceae recorded the highest values during the winter season almost in most of the sites, which represent the color on the map .While bing the highest value Cyanophyceae wase 172.37×10^4 cell/gm While Frugilarophyceae the values were few in all locations and Coscinodiscophyceae disappeared during the winter season also in all locations, so they appeared on the map in one color .

In winter 2019 in the same site this result agrees with [22, 23] and that is owing to that the rainfall drifts organic materials resulted from the fertilizers which contain phosphorus and nitrogen that are an important nutrient for the algae from the nearby agriculture area in Site 4 [24], or it may return to the increase of the organic materials in the water column because the low temperatures decrease the decomposition of these materials, also the increasing of silicate concentration in winter may be from the contributed factors that causing the recorded increasing [25].

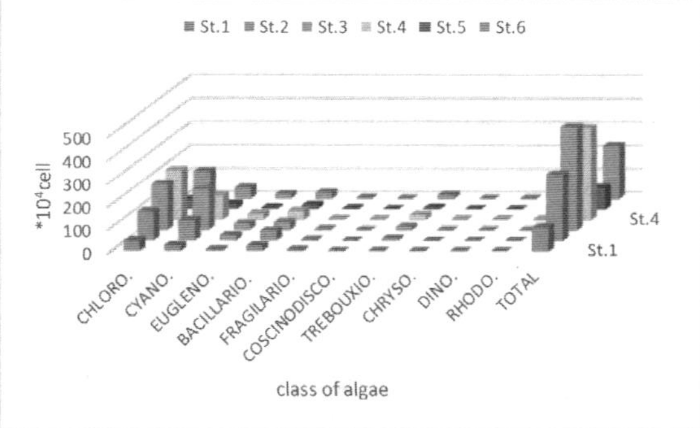

IOP Conf. Series: Earth and Environmental Science 1002 (2022) 012012 doi:10.1088/1755-1315/1002/1/012012

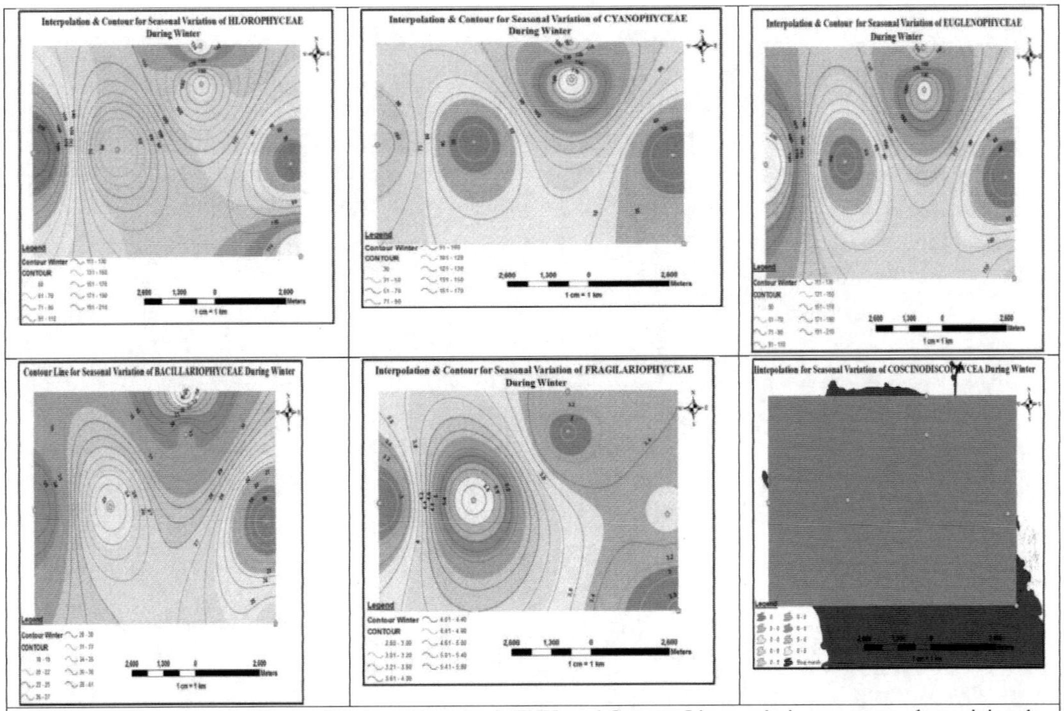

Fig 3. The interpolation of Inverse Distance Weighted (IDW) and Contour Lines techniques process determining the quantities.

Bacillarophycaea recorded the highest values during the spring season 448.74× 10^4 in most of the sites. While Euglenophycaea recorded lowest value 10.1× 10^4 on the spring season. The spring season showed the lowest values 811.2× 10^4 for class densities in all seasonal.(figur4).Can be determined through this modern technology the expected areas or predict the density of algae in the area bounded between two sites and is considered a good area of interest.

Total algae peaked (161.11 × 10^4 cell/gm) in the spring of 2019. This corresponds to the study of southern Iraq waters [26] . The increase in the number of algae attached to the *P. australis* plant may also be explained by the exchange of the existing plant nutrients. Between this plant and the attached algae, as these algae grow on the organic materials that are secreted by the plant significantly at the beginning of the growing season (spring) and with the passage of time the surface of the plant (leaf or stem) becomes covered with a dense layer of different organisms such as fungi and bacteria that may compete Algae are attached to the nutrients, so the number of algae decreases towards the summer season coinciding with the decrease in nutrient concentrations [27]. The decline in spring may occur because of that the juvenile plants of P.australis are resistant against the secreted attaching substances by algae which aid to strengthen their attach or it may return to that the emergent leaves of *P.australis* widely affect by the variations in the surface water and wave movements [28].

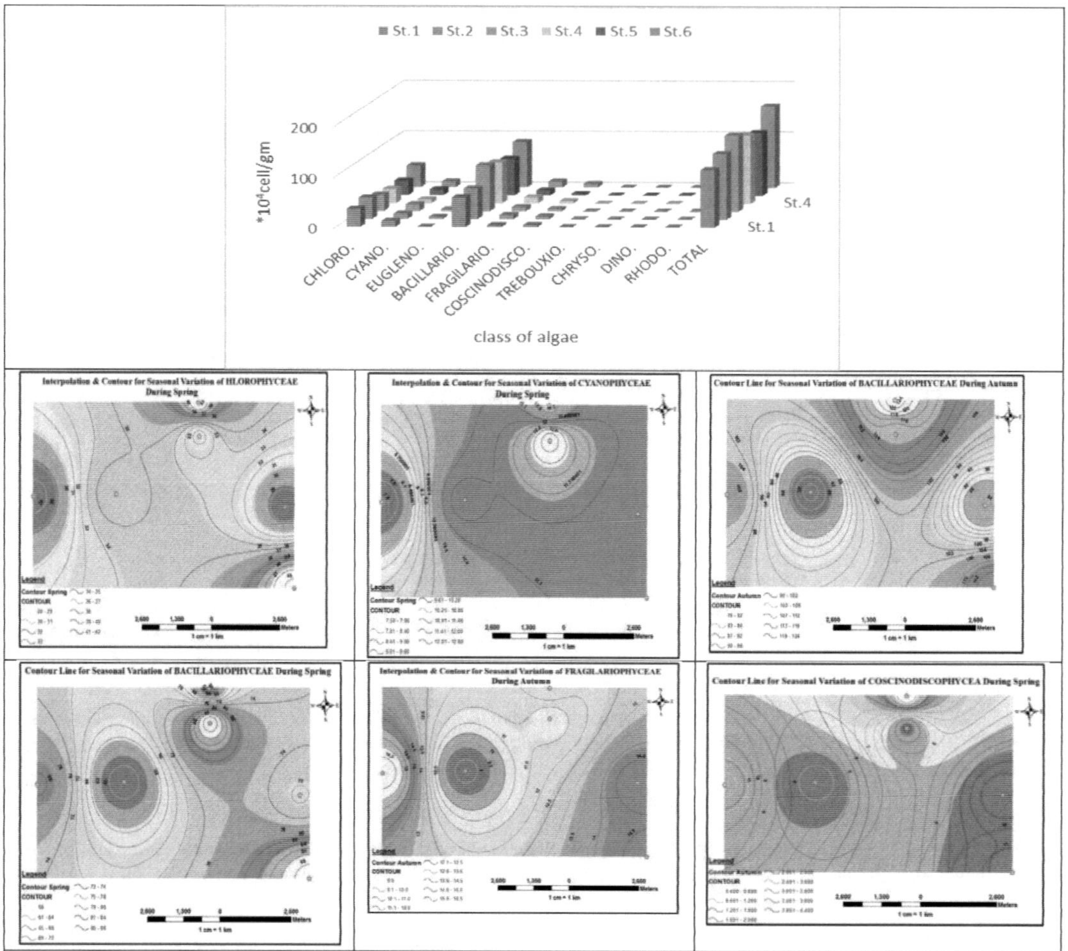

Fig 4. The interpolation of Inverse Distance Weighted (IDW) and Contour Lines techniques process determining the quantities.

Bacillarophyceae were recorded highest values during all seasonal of the study as well as during all sites1728.72× 10^4 cell/gm superior to all classes .while Euglenophyceae were lowest value all the site 26.79×10^4cell/gm.This is evident in the contour maps through the different colors and contour lines between the sites In general Bacillariophyceae recorded high total number of epiphytic algae, followed by Chlorophyceae that was more than Cyanophyceae,

The results of the study also showed a seasonal variation in the quantitative composition of adherent algae on aquatic plants, as the summer of 2019 was characterized by being the highest in the number of adherent algae species recorded in the other seasons of the study.

This may be related to the increase in temperature, the intensity of lighting, the length of daylight hours and the availability of many materials.

The decomposition of organic matter or resulting the metabolic activities of aquatic plants, which is an important source of many plant nutrients (nitrogen and phosphorous) which are consumed by algae [29].this study agree to study [30] in the Hawizeh marshes.

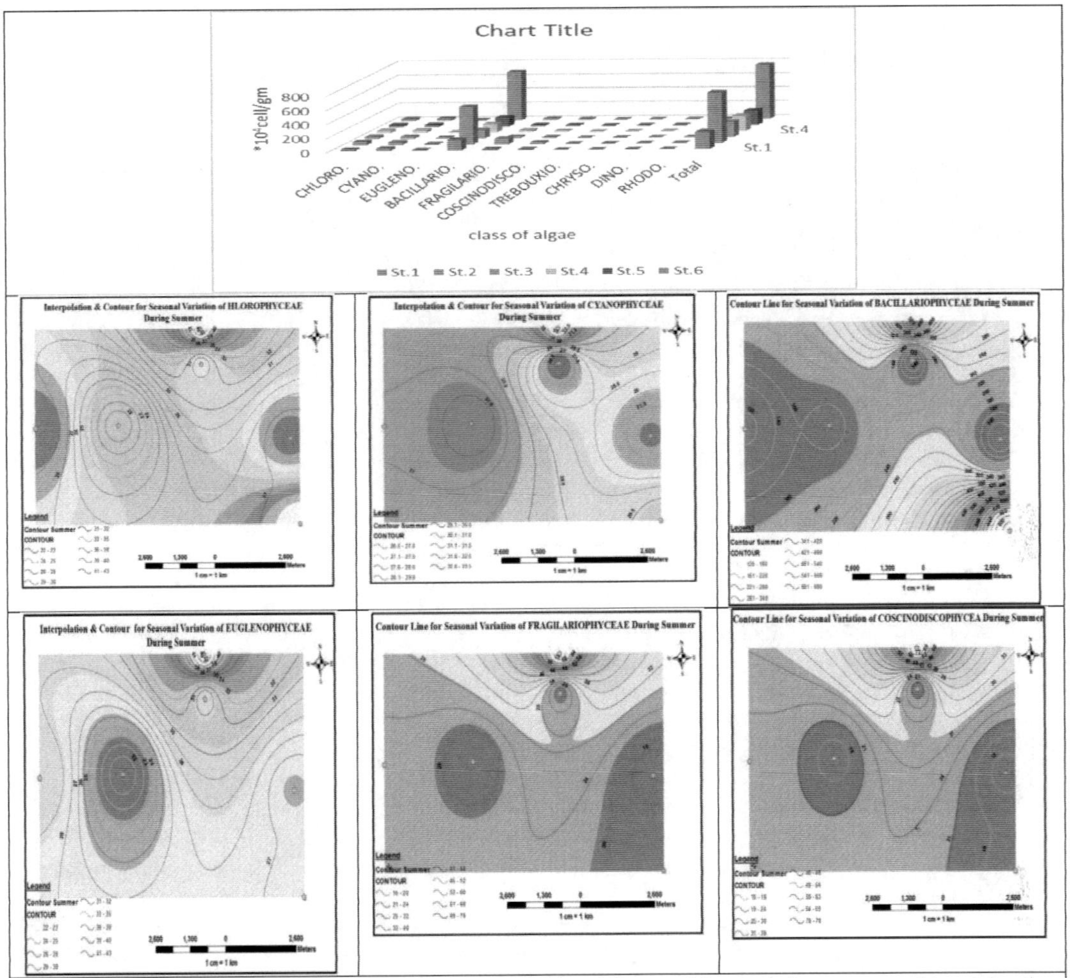

Fig 5. The interpolation of Inverse Distance Weighted (IDW) and Contour Lines techniques process determining the quantities.

4. Conclusions

1- maps were produced to predict the unknown, distant and hard-to-reach values surrounding the sites from which samples were taken, which are very suitable for knowing good and polluted water quality. species in the Um Al-Naaj Lake. Through digital maps of statistical interpolation technology and imaginary contour lines, each color is considered to be the true and expected value of the factors affecting water quality without reaching this area.

2- The seasonal variation in the quantitative composition of epiphytic algae was characterized by the summer of 2019 having the highest number of attachment algae species recorded in the other seasons of the study.

3- The total number of epiphytic algae on the aquatic plant *Phragmites australis* during the study period 2018-2019 was 306 species and belonged to 61 genera and species of class Bacillariophceae species were predominant upon other species, where it recorded 215 species.

References

[1] **Garstecki T, Amr Z. 2011.**Biodiversity and ecosystem management in the Iraqi Marshlands–screening study on potential World Heritage Nomination. IUCN (International Union for Conservation of Nature), Amman, Jordan.:55-7.

[2] **Fouad K. Mashi, 2018.**Monitoring Al-Hammar Marsh Topography and Climatic Applied Satellied MODIS Imagery, Indian Journal of Natural Sciences, Vol. 8, Issue 47, pp: 13705-13714,

[3] **Fouad K. Mashee Al Ramahi, Zina Khalil Ibrahim Al Bahadly. (2020).** The Spatial Analysis for Bassia eriophora (Schrad.) Asch. Plant Distributed in all IRAQ by Using RS & GIS Techniques. Baghdad Science Journal: 17(1): 126-135.

[4] **Hassan FM, Hadi RA, Kassim TI, Al-Hassany JS.2012.** Systematic study of epiphytic algal after restoration of Al-Hawizah marshes, southern of Iraq. Int. J. of Aquatic Science. Jan 1;3(1):37-57.

[5] **Dunn,A.E; Dobberfuhl D R; Casamatta,D.A.(2008).**A survey of algal epiphytes from Vallisneria Americana Michx. (Hydrocharitaceae) in the lower St. Johns River, Florida. Southeastern Naturalist, 7(2): 229–244.

[6] **Fouad K. Mashee Al Ramahi1, Zina Khalil Ibrahim Al Bahadly, 2017**, "Estimation of Suaeda aegyptiaca Plant distribution regions at Iraq using RS & GIS Applications", Iraqi Journal of Science, Vol. 58, Issue. 2A, pp: 767-777, DOI:10.24996.ijs.2017.58.2A.20.

[7] **Fawzy M, Nasr M, Adel S, Nagy H, Helmi S .2016.** Environmental approach and artificial intelligence for Ni (II) and Cd (II) biosorption from aqueous solution using Typha domingensis biomass. Ecological Engineering. Oct 1;95:743-52.

[8] **Neran A. AL Naqeeb, Jinan S. Al Hassany, Fouad K. Mashi. (2020).** Assessment of the water quality of Um El-Naaj Marshes by Diatoms. Eco. Env. & Cons. 26 (1); 449-464.

[9] **Hassan FM, Al-Bdulameer SH.2014**. Qualitative and Quantitative Study of Epipelic Algae in Tigris River within Baghdad City, Iraq. J Baghdad Sci.;11:1074–82.

[10] **Neran A. AL Naqeeb, Jinan S. Al Hassany, Fouad K. Mashi. (2020).** Assessment of the water quality of Um El-Naaj Marshes by Diatoms. Eco. Env. & Cons. 26 (1); 449-464.

[11] **Al-Hassny JS, Al-Bueajee 2015**. A Qualitative study of epiphytic algae (diatom) on some aquatic plants in Al-Auda marshes within Maysan Province/Southern Iraq. Baghdad Sci J. 2015;12(4):665-76.

[12] **Israa J. Muhsin, Fouad K. Mashe, Rafid J. Tawfeeq, 2011.** "Monitoring the Vegetation and Water Content of AlHammar Marsh Using Remote Sensing Techniques", , Baghdad Science Journal, Vol.8(2), pp: 646-651.

[13] **USGS (United States Geological Survey), (2017).**Landsat—A Global Land-Imaging Mission. [Online] Available at: http://remotesensing.usgs.gov.

[14] **Fouad K. Mashee, Osamah Hadi Mutlag, Muaid Jassim Rasheed. 2020.** Spectral indices analysis for Al-Gharaf River basin land cover by apply remote sensing techniques. Eursian Journal of BioSciences; 14(2): 3367-3375.

[15] **Israa J Muhsin, K Mashee Foud, 2012,** "Improving Spatial Resolution Of Satellite Image Using Data Fusion Method, Iraqi Journal of Science: 53(4); 943-949.

[16] **Israa J.Muhsin, Faleh Hassan Mahmood and Fouad K. Mashee, 2012**, "Multi Spectral Scanner "MSS" and Panchromatic Components Difference of Al- Haditha Dam Region Using GIS and Remote Sensing Techniques", Al-Mustansiriyah J. Sci: 23(6); 233-242.

[17] **Fouad K. Mashee Al Ramahi, Mazin Shakir Jasim, Muaid Jassim Rasheed. 2020.** To study climatic factors effect on Land Covers (LC) for Salah Aldeen region by using remote sensing data. ECOLOGY, ENVIRONMENT AND CONSERVATION; 26(1): 446-453.

[18] **Peter, A. B. and Rachael, A. M. 1998.** Principles of Geographical Information Systems. Oxford University Press Inc., New York, 17-34.

[19] **Jasim, Mazin S.1; MASHEE, Fouad K. 2020**. Monitoring and calculating the carbon emissions in Baghdad and its effect on increasing temperatures from 2003-2018 using remote sensing techniques data. Periódico Tchê Química: 17 (n°37); 117-128.

[20] **Abdulrahman B. A1, Fouad K. Mashee and Sabah N. Kadim, 2018**. Study Spectral Indices of Land Cover Around Al-Shari Lake and Produce Mapping by Using Remote Sensing Technique, Indian Journal of Natural Sciences, Vol.8 , Issue 49, pp: 14215-14224,

[21] **AL-Saeady RN.2016**. An Ecological study of epiphytic algae on Aquatic Macrophytes in Tigris River within Baghdad city/Iraq (Doctoral dissertation, M. Sc. Thesis. College of Science for Women, University of Baghdad. pp: 165).

[22] **Hindi MT. A** Study of Epiphytic and Epipelic Algae in Al-Dora Site / Tigris River in Bagdad Province- Iraq Introduction : Materials and Methods : Results and Discussion : Baghdad Sci J.;13(4):721–33.

[23] **Ali K. Mohammed Ali, Fouad K. Mashee Al Ramahi, Abdul-Rahman B. Ali. (2019).** Evaluation of impact of vegetation decrease on precipitation rates in Baghdad City using remote sensing technique. Eco. Env. & Cons: 25 (November Suppl. Issue); S48-S54.

[24] **Abdul IM ,2010**.. Evaluation of the Environment of Hor El-Gabayish by the adoption of environmental and life evidence. Collage of Agriculture, University of Basra,;

[25] **Muaid Jassim Rasheed and Fouad K. Mashee Al-Ramahi. 2021**. DETECTION OF THE IMPACT OF CLIMATE CHANGE ON DESERTIFICATION AND SAND DUNES FORMATION EAST OF THE TIGRIS RIVER IN SALAH AL-DIN GOVERNORATE USING REMOTE SENSING TECHNICQES. Iraqi Geological Journal; 54(1A): 13-23.

[26] **Al-Farhan SN 2010.** An ecological study of the benthic aquatic ecosystems of Basrah. Ms. C. thesis, University of Basrah, Iraq;

[27] **Fouad K. Mashee, Gheidaa S. Hadi, 2017,** "Study the Wet Region in Anbar Province by Use Remote Sensing (RS) and Geographic Information System (GIS) Techniques", Iraqi Journal of Science, Vol. 58, Issue. 3A, pp :1333-1344, DOI: 10.24996/ijs.2017.58.3A. 18.

[28] **Fouad K. Mashee Al Ramahi , Muaid Jassim Rasheed and Mazin Shakir Jasim. 2020.** The determination of radon gas concenetration distributed North of Baghdad Governorate: spatial analysis by applying remote Sensing (RS) and geographic information system (GIS) techniques. Eco. Env. & Cons. 26 (3); 26(3): 987-993.

[29] **Taif Adil Dhamin, Fouad K. Mashee, Ebtesam F. Khanjer. 2020.** Detection Agriculture Degradation for the South of Baghdad City Using Remote Sensing Data for Years 2010-2019. Iraqi Journal of Science: 2(4); 57-66.

[30] **Al-Hassany JS**. *A study of the ecology and diversity of epiphytic algae on some aquatic plant in Al-Hawizah marshes, Southern Iraq* (Doctoral dissertation, Ph. D. thesis, College of science for women, University of Baghdad. pp: 1-215 (in Arabic)).

The effect of climate change on energy consumption and environmental pollution according to degree days indicator "Baghdad city as an example"

Susan Abed Hassan[1], Saad Ismail Husain[2]

[1] Department of Architectural engineering, College of Engineering, Al- Nahrain University-Iraq

[2] Renewable Energy and Environment Research Center, Ministry of Industry and Minerals, Iraq

[1]Email: dr.susanabedhassan@eng.nahrainuniv.edu.iq

Abstract: The problem of climate change has emerged globally in recent decades. One of its most prominent indicators is the increase in temperatures in all countries, especially in regions with hot and arid climates. This has coincided with the problem of rising energy consumption for cooling needs in cities, especially in the summer season. Also, the increase in energy consumption contributes to increasing environmental pollution in urban areas. This, in turn, contributes to global warming. Several methods have emerged to calculate the energy consumption of buildings. The degree-days method is more applicable because it provides an accurate analysis of the temperature of the weather and its change on any given day and throughout the seasons of the year. This research calculated the cooling degree days for Baghdad City from 1960 until 2019. The results revealed that the annual average of degree days measured over the years had increased by 24% since 1960; energy consumption was calculated accordingly to determine the energy requirements, which had increased by 30%. The calculation of degree days will enable a local climate to assess critical issues of climate change and energy consumption related to environmental pollution.
Keywords: Degree days, Energy consumption, environmental pollution

1. Introduction

There is a global problem of climate change that affects all cities worldwide, especially in countries close to the equator due to the nature of their hot, arid climate, lack of precipitation, and wind fluctuations. Climate change is defined as (the change in the general climatic conditions resulting from human activities, directly or indirectly). The first definition of climate change dates back to 1824 by the scientist Fourier, who described the greenhouse phenomenon as the retention of gases resulting from the fuel used by humans in the Earth's atmosphere, generating an increase in the temperatures recorded annually and the associated rise in sea levels and climate fluctuations. The primary indicator of climate change is rising temperatures in cities around the world; average recorded temperatures have risen by 1.5 degrees Celsius globally [1]. An increase in buildings' energy consumption for cooling needs in hot-climate cities. Also, energy production contributes to increased pollution of the environment (air, water, thermal, etc.). As a result, global warming is exacerbated. [2]. It starts from its early stages of design and construction and the energy consumed by the construction activities themselves [3]. The energy consumption for buildings during occupancy is estimated at 40% of the total energy in the cities [4]. One of the methods used to calculate the energy consumption of buildings is the Degree-days method. It began in agricultural research, where knowledge of the cumulative variation in outdoor air temperature is important. The concept of the degree of the day was adopted in the calculations of the thermal loads of buildings by calculating the differences between the external and internal temperatures and their rates during the different months of the year [5]. Previous research from around the world yielded a plethora of papers on various aspects of climate change and degree-day calculations [6, 7, 8, 9]. And in Iraq, many researchers studied the climate change evidence and degree day's analyses [10, 11 & 12]. But the stud. However, the statistical study of the effect of climate change on energy consumption and environmental pollution, using the degree day indicator for Baghdad city as an example, has never been done before in a statistical way.

2. Energy consumption and pollution to the environment

The consumption of fossil fuel represents the basis of contemporary life. All industries, transportation, and daily services consume large amounts of energy. It leads to great environmental pollution associated with all stages of energy achievement, starting from obtaining sources from the ground and treating them to obtain their various types and then transferring them to generating stations. In all these stages, the pollution problems of the surrounding environment arise from the pollution of air, water, and land. Pollution problems increase near rivers and water bodies since most power plants are located near rivers because of their production operations. The leakage of fuel into the ground pollutes the groundwater. This pollution reaches drinking water sources as well. The increase in energy consumption in cities is related to the increase in pollution. This reduces the requirements for a healthy life for its residents. Therefore, it is necessary to present proposals and alternatives to reduce energy expenditure and rely on clean energy alternatives. [13,14].

3. Climate change in Iraq

Iraq is located in the Middle East and North Africa. The capital Baghdad is located at 33° 19′ N, 44° 25′ E.Its climate is classified as hot and dry in summer, cold and humid in winter. Summer peak temperatures reach higher than 50°C when the average max temperatures in summer reach 45°C. Climate changes are evident in Iraq; the climatic elements have changed in different means since 1950, including increasing yearly temperatures by 0.7 °C in each era. Also, there was a shortage in annual rainfall at 1.3 - 6.2 mm/year; there was an increase in dust storms in all parts of Iraq [2].

4. Materials and methods

The main objective of this research is to find climate change effect on energy consumption according to degree day's indicator in Baghdad hot arid climate through:
-Studying and calculating the historical degree of days in the city of Baghdad from 1960 to 2019.
-Study and analyze energy expenditure in the model of walls and ceilings adopted in residential buildings according to the degree of days factor in Baghdad city, with the area of (100) m2.
The following formulae were used if the minimum temperature was larger than the base temperature [6]:

$$CDD = \frac{(T_{max}+T_{min})}{2} - T_{base} \qquad (1)$$

1. The following formulae were used if the temperature is equal to or smaller than the base temperature:

$$CDD = \frac{(T_{max}+T_{min})}{2} - \frac{T_{base}-T_{min}}{4} \qquad (2)$$

2. The formula that relates the cooling degree days and energy consumption for cooling is:

$$Cooling\ Energy\ Lost = (24 \times CDD\ Annual \times U)/1000\ \text{kWh/year/m}^2 \quad (3)$$

5. Results

5.1 Cooling Degree Days for Baghdad city

A calculation was made for the cooling degree days for Baghdad city and the base temperature (21°C) for the periods of time (1960-2019) are shown in Figure (1). It's obvious that cooling degree days are higher for recent years. The degree days for Baghdad city increased by about 24% form the year 1960.

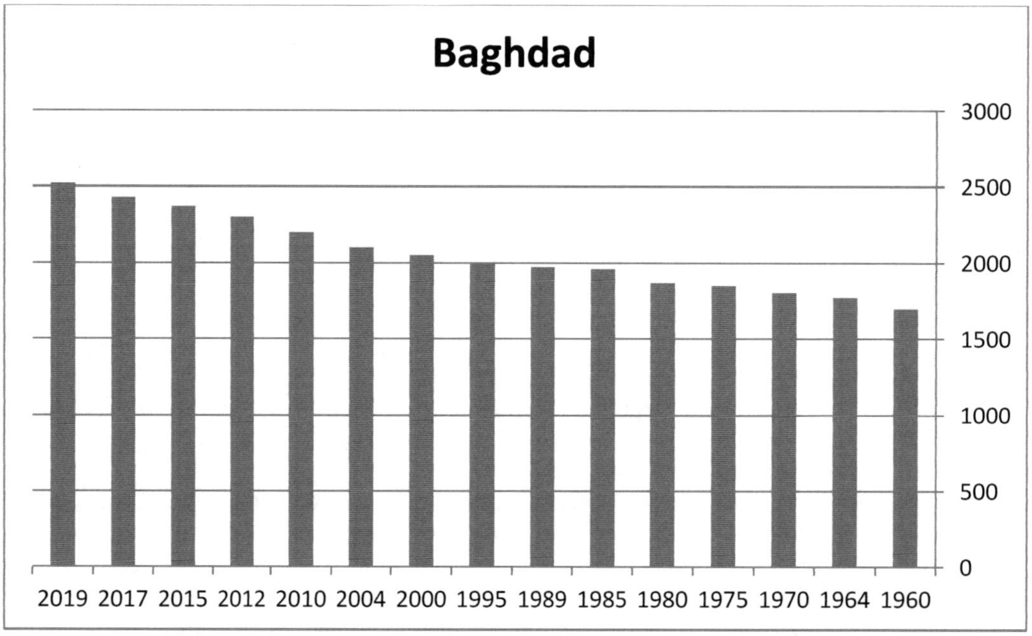

Figure 1. Average cooling degree days for Baghdad city (1960-2019)

5.2 Energy consumption comparison for the residential buildings

Baghdad city's standard residential building materials are from clay brick with (1.429) U-Value. And for the ceiling is the concrete cast with precast concrete tile and river sands that have (1.1696) U-Value. The energy consumption changes between(1960-2019) are shown in Table 1 and Fig 2 & 3. The increase in energy consumption in kWh/year/m2 reached 32%

Table 1. The energy consumption in kWh/year/m^2 for walls and ceiling

The year	Energy consumption kWh/year/m^2	
	Walls	Ceiling
1960	41.75472	51.0153
1964	43.47403	53.11593
1970	44.21088	54.0162
1975	45.43896	55.51665
1980	45.93019	56.11683
1985	48.14074	58.81764
1989	48.38635	59.11773
1995	49.1232	60.018
2000	50.35128	61.51845
2004	51.57936	63.0189
2010	54.03552	66.0198
2012	56.49168	69.0207

2015	58.21099	71.12133
2017	59.68469	72.92187
2019	61.89523	75.62268

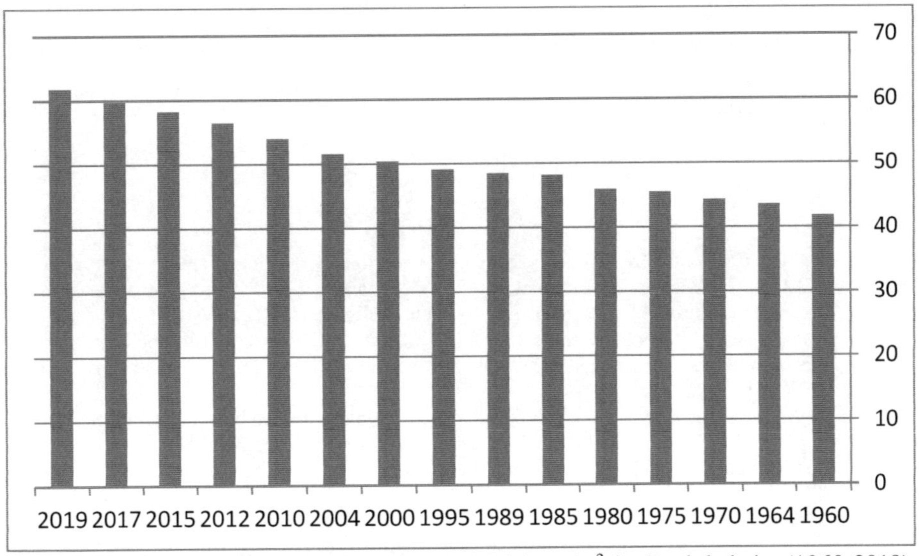

Figure 2 . The walls energy consumption in kWh/year/m^2 for Baghdad city (1960-2019)

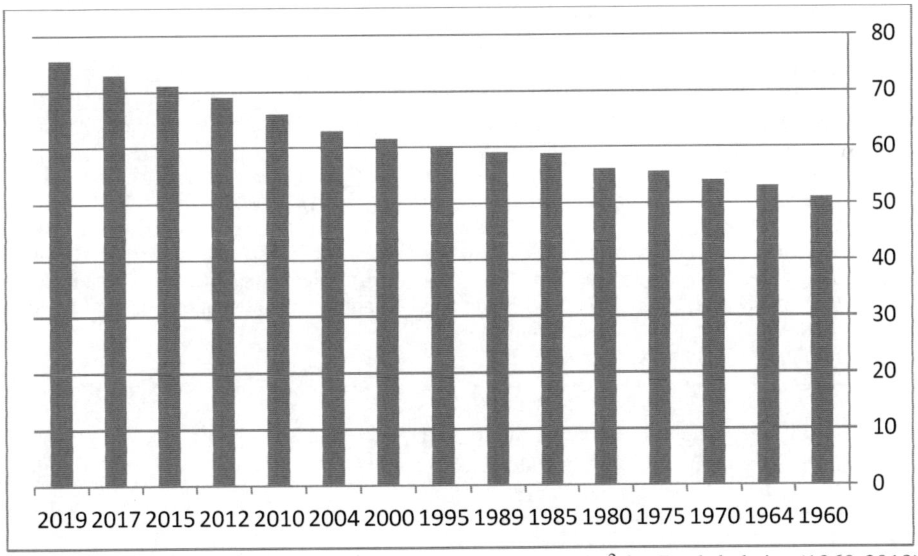

Figure 3 . The ceilings energy consumption in kWh/year/m^2 for Baghdad city (1960-2019)

6. Conclusion

Baghdad City is suffering from the impacts of climate change. There was an increase in temperature that caused the city to warm. The temperatures recorded from 1960 showed an increase in annual temperature rates that exceeded previous years. This increase has affected the increased demand for energy consumption for buildings. Also, energy consumption is related to environmental pollution (air, water, thermal, etc.) in urban areas. The energy consumption of traditional homes in 1960 was no longer sufficient to suit the year 2019, which contributed to the increase in electrical energy expenditures. An examination of cool degree days in Baghdad City from 1960 to 2019 showed that degree days for Baghdad City increased from 1960 to 2019) by about 24%. Then energy consumption for the buildings of the typical house (walls and ceilings) was calculated accordingly. The increase in energy consumption in kWh/year/m2 reached 32%, which means that the same residential building from the 1960s needs to increase the cooling energy by 30% to achieve thermal comfort for the occupants now.

Reference

[1] United Nations Framework Convention on Climate Change: Convention. UNFCC. Article 1 –Paragraph 2, 1994
[2] Susan Abed Hasan.(2018). The impact of residential building's design on the energy consumption in hot desert climate (Baghdad city as an example). Journal of Urban and Environmental Engineering 12, no. 1: 88-92.
[3] Susan Abed Hassan, Jamal Abed al Wahid Jassim.(2019). The role of multi-story structural building systems on reducing embodied energy consumption and carbon emissions. IOP Conf. Series: Materials Science and Engineering 518, 022031. doi:10.1088/1757-899X/518/2/022031.
[4] Alsammarae Aiham. (2005). Iraq's electrical system status and future plan Electric power in Chicago.
[5] Susan Abed Hassan.(2019). The Role of Cooling Degree Days on determining the insulation of building Envelop in a hot climate (Iraqi cities as an example). IOP Conf. Series: Materials Science and Engineering 584, 012009. IOP publishing doi:10.1088/1757-899X/584/1/012009
[6] Moustris K P P T, Zacharia I K, Larissi P T Nastos, and A G Paliatsos. (2011). Cooling and heating degree-days calculation for representative locations within the greater Athens area, Greece. Proc 12th Conf on Environmental Science and Technology Rhodes Greece Global NEST A1279–A1286.
[7] Day Tony Franklin Bryan Fry Martin Holme Michael (2006). Degree-days theory and application. The Chartered Institution of Building Services Engineers 222 Balham High Road, London SW12 9BS http://trade.gov/static/iraq_electricalppt.pdf
[8] Carbon Trust. (2007). Degree days for energy management A practical introduction Carbon Trust, pp. 20.
[9] Boujnah Mohamed Jraida Kaoutar Farchi Abdelmajid Mounir Ilham.(2016).Comparison of the calculation methods of heating and cooling degree-days in three different cities in morocco. International Journal of Current Trends in Engineering & Technology ISSN: 2395-3152. Volume: 02, Issue: 06
[10] Al-Waeli, A., Al-Asadi, K. and Fazleena, M. (2017). The impact of Iraq climate condition on the use of solar energy applications in Iraq: a review. International Journal of Science and Engineering Investigations, 6(68): 64-73
[11] Mohammed M.Ahmed,Omar L.Khaleed, NaghamT. Ibraheem. (2017). Calculation of Daily Cooling Degree for Selected Stations in Iraq. Diyala journal for pure science, Vol 14 No: 4, October 2018. http://dx.doi.org/10.24237/djps.1404.440C
[12] A. A. Alaskary, A. M. Hasson, M. J. Jweeg, M. L. Al-Waily. (2020). Microclimate Energy Considerations in Building Design for Arid Regions. Nature Environment and Pollution. Technology an International Quarterly Scientific Journal V19 N 3, https://doi.org/10.46488/NEPT.2020.v19i03.024
[13] https://www.eionet.europa.eu/gemet/ar/concept/2833
[14] ".التميمى، بشير فرحان محمود، و عبدالرحمن بن عبدالكريم اليحيى(2018). تلوث مياه شط العرب بالعناصر الثقيلة الناتجة عن محطات الطاقة الحرارية وتأثيراتها الصحية .مجلة آداب الفراهيدي: جامعة تكريت - كلية الآداب مج10, ع34: 179 - 193
[15] Climate Change Profile Iraq, Ministry of Foreign Affairs of the Netherlands |(April 2018), P.O. Box 20061 | 2500 EB The Hague | the Netherlands www.government.nl/foreign-policy-evaluations

AUTHOR INDEX

Abaas, Manar A. 10
Al Hassany Jinan S. 154
Al Naqeeb Neran A. 154
Al-Hamadany, Qusay Hamid 19
Ali, Ali Abdulkhabeer 113
Aljaafari, Raad K. 10
Al-Sheikhly, Omar F. 10
Al-Taai, Suaad Hadi Hassan 89
Al-Tufaily, Hodoud Mohamed Abboud 121
Al-Zurfi, Sienaa M. 10
Aswad, Zahraa S. 30
Boden, Rich 65
Braungardt, Charlotte 65
Fazaa, Nadheer A. 10
Haba, Mukhtar K. 10
Hannun, Rafid M. 104
Hashim, Athraa 139
Hassan, Susan Abed 164
Husain, Saad Ismail 164
Huseen, Basim 139
Ibrahim, Lamia Flaieh 121
Irzoqy, Israa Mahmooed Mohammed 121
Jasim, N A 1
Jebril, Nadia 65
Kareem, Ali 10
Kazm, Karrar H. 51
Mashee Fouad K. 154
Mizhir, A H 1
Mohammed, Rusol 139
Najim, Sarmad T. 51
Nasir, Noori Abdul-Nabi 19
Rahi, Khayyun 113
Razzaq, Ali H. Abdul 104
Salman, Havan H. 30
Whittingham, Mark J. 10
Yesser, Abdulkareem Taher 19

Institute of Physics
Dirac House, Temple Back
Bristol BS1 6BE UK

ISSN: 1755-1307
ISBN 978-1-7138-5444-9

CURRAN ASSOCIATES INC.
proceedings.com